I want to dedicate this work to my Russian scientific supervisor Garald Isidorovich Natanson (May 9, 1930–July 24, 2003), professor at the chair of mathematical analysis of St Petersburg State University, доктор of mathematical sciences.

Thank you, Garald Isidorovich, that I was allowed to meet you and listen to your vivid narrations about what mathematics in St Petersburg was about. I hope that I could save just a little bit of your spirit and so fill this work with life. R.I.P.

Consulting Editor

George A. Anastassiou
Department of Mathematical Sciences
University of Memphis

Karl-Georg Steffens

The History of Approximation Theory

From Euler to Bernstein

Birkhäuser
Boston • Basel • Berlin

Karl-Georg Steffens
Auf der neuen Ahr 18
52372 Kreuzau
Germany
kgsteffens@web.de

Consulting Editor:
George A. Anastassiou
University of Memphis
Department of Mathematical Sciences
Memphis, TB 38152
USA

Cover design by Mary Burgess.

AMS Subject Classification: 11J68, 32E30, 33F05, 37Mxx, 40-XX, 41-XX, 41A10

Library of Congress Control Number:

ISBN-10 0-8176-4353-2   eISBN 0-8176-4353-9
ISBN-13 978-0-8176-4353-9

Birkhäuser

(TXQ/SB)

9 8 7 6 5 4 3 2 1

www.birkhauser.com

Preface

The aim of the present work is to describe the early development of approximation theory. We set as an endpoint the year 1919 when de la Vallée-Poussin published his lectures [Val19]. With these lectures all fundamental questions, that is, non-quantitative theorems, series expansions and quantitative problems, received their first summarized discussion.

The clear priority of the present investigations are the contributions of Pafnuti Lvovich Chebyshev and of the St Petersburg Mathematical School founded by him. Although some overviews and historical contributions have been published on this subject (e. g., [Gon45], [Gus61] and [But92]) we think that nevertheless it makes sense to go into this topic again for at least five reasons:

Firstly, you find contradictory statements about the exact efforts of Chebyshev and his pupils. So the statement that Chebyshev himself proved the alternation theorem is wrong and the claim that St Petersburg mathematicians had not been interested in the theory of functions is pure nonsense.

Secondly, the available material of Soviet origin is sometimes tendentious and exaggerated in its appreciations of the persons involved, both positively in the almost cultic adulation of Chebyshev and negatively in neglecting the scientific results of mathematicians like Sochocki and his students who did not stand in the limelight, or in belittling the work of Felix Klein.

Thirdly, nearly all historical comments are written in Russian or in one of the languages of the Soviet Union (except for some articles, for example the contributions of Butzer and Jongmans ([BuJo89], [BuJo91] and [BuJo99]) and some papers of Sheynin. So it was time to explain this era of enormous importance for the development of mathematics in Russia and the Soviet Union to those who are not able to read Russian and do not have the time or opportunity to dig in Russian archives and libraries.

In this regard, we feel that it would be disrespectful and unhelpful to refer to Russian contributions which nearly no-one could have access to. Therefore you will find many quotations from the works listed in the References.

Fourthly you will recognize that we did not want only to *describe* the results of the St Petersburg Mathematical School, but also to discuss its historico-philosophical background, and so its character and how it interacted with other European schools.

And lastly we present some interesting facts about the rôle played by Göttingen in spreading Russian contributions and in their further development.

The breadth of this subject made some restrictions necessary. Definitely you will miss the problems of moments. But we think that also without them the basic tendency of the development described here would not have changed. Only the rôle of A. Markov, Sr. then would have been of even higher importance.

We did not analyse the work of Bernstein as carefully as the contributions of other authors because the German translation of Akhiezer's scientific biography of Bernstein [Akh55] was published in 2000 by R. Kovacheva and H. Gonska.

Expression of Thanks

I especially want to express my deep gratitude to George Anastassiou (Memphis, TN) and Heiner Gonska (Duisburg, Germany). Only by their initiative could this work be published in its present form. Many thanks to Birkhäuser that it followed their recommendation.

Also I want to thank all the people who helped me when I collected material, did not have the necessary ideas or asked for an English word. Special thanks to the wonderful support I enjoyed in St Petersburg and Kyiv, especially from Garald Isidorovich Natanson, Nataliya Sergeevna Ermolaeva and Aleksey Nikolaevich Bogolyubov.

And finally I will never forget Bruno Brosowski (Frankfurt/Main, Germany) who initially led me to the fantastic Chebyshev approximation theory.

Kreuzau, October 2004   *Karl-Georg Steffens*

Introduction

"All exact science is dominated by the idea of approximation."

This statement, attributed to Betrand Russell, shows the borders of exact science, but is also intended to point out how to describe nature by means of mathematics.

The strength of mathematics is abstraction concentrating on simple and clear structures which it aims to rule completely.

To make abstract theories useful it is necessary to adopt them to certain *a posteriori* assumptions coming from reality: Measured data cannot be more exact than the instrument which recorded them; numerical computations are not better than the exactness of the computer.

Whenever one computes, one *approximates*.

It is therefore not surprising that the problem of approximative determination of a given quantity is one of the oldest challenges of mathematics. At least since the discovery of irrationality, considerations of this kind had become necessary. The formula for approximating the square root of a number, usually attributed to the Babylonians, is a case in point.

However, approximation theory is a relatively young mathematical branch, for it needed a concept which describes the mutual dependence between quantities exactly, i.e., the concept of a function. As is well-known, the first approach to defining a function based on this dependence and to abstract from formulae was developed by Euler.

The first abstract definitions of this concept were followed by reflections on how to represent functions to render them pratically useful. Thus, formulae were developed to assist in approximating mainly transcendental functions. At first these representations relied on Taylor's formula and some interpolation formulae based on Newton's ideas.

Although these formulae gave good approximations in certain special cases, in general they failed to control the approximation error because the functions were not approximated uniformly; the error grew beyond the interpolation points or the expansion point. The least square method developed by Gauß provided some improvement, but points might still exist within the interval considered where the error of approximation becomes arbitrarily large.

It follows that new ideas had to be found to solve problems in those cases where it was important to control approximating errors over whole intervals. The present work starts with the first known problems which made such considerations indispensable.

Probably the first work on this subject is attributed to Leonhard Euler who tried to solve the problem of drawing a map of the Russian empire with exact latitudes. He gave a best possible approximation of the relationship between latitudes and altitudes considering all points of one meridian between given latitudes, i. e., over a whole interval [Eul77].

Because of the enormous size of the Russian empire all known projections had very large errors near the borders of the map, therefore Euler's approach proved helpful.

A problem encountered by Laplace was similar in character. One paragraph of his famous work [Lap43] (first published in 1799) dealt with the question of determining the ellipsoid best approximating the surface of the earth. Here, too, it was important to have the error held small for every point on the earth's surface.

Euler solved his problem for a whole interval, whereas Laplace assumed a finite number of data which was very much larger than the number of parametres in the problem. This fact alone prevented a solution of the problem by interpolation.

In 1820 Fourier generalized Laplace's results in his work 'Analyse des équations determinés' [Fou31], where he approximatively solved linear equational systems with more equations than parametres by minimizing the maximum error of every equation.

In 1853 Pafnuti Lvovich Chebyshev was the first to consolidate these considerations into the 'Theory of functions deviating the least possible from zero,' as he called it.

Starting out from the problem of determining the parameters of the driving mechanism of steam engines—also called Watts parallelogram—in such a manner that the conversion of straight into circular movement becomes as exact as possible everywhere, he was led to the general problem of the uniform approximation of a real analytic function by polynomials of a given degree. The first goal he achieved was the determination of the polynomial of nth degree with given first coefficient which deviates as little as possible from zero over the interval $[-1, 1]$. Today this polynomial is called a Chebyshevian polynomial of the first kind.

Further results were presented by Chebyshev in his work 'Sur les questions des minimas' [Cheb59] (written in 1857), where he stated a very general problem: that of determining parameters $p_1, \ldots, p_n$ of a real-valued function $F(x, p_1, \ldots, p_n)$ so that over a given interval $[a, b]$ the maximum error

$$\max_{x \in [a,b]} |F(x, p_1, \ldots, p_n)|$$

is minimized. Under certain assumptions for the partial derivatives

$$\frac{\partial F}{\partial p_1 \ldots \partial p_n}(x, p_1, \ldots, p_n)$$

he was able to prove a generally necessary condition for the solution of the problem.

Using this condition he showed that in special cases (polynomial, weighted polynomial and rational approximation) it led to the necessary condition that F has a fixed number of points where it assumes the maximum value. These points are now known as deviation points. However, the alternation theorem which clearly follows from this result has never been proved by Chebyshev himself.

The aim he sought to achieve with this contribution is to find the polynomial uniformly deviating as little as possible from zero for any number of given coefficients. Later his pupils would work on several problems arising from this general challenge. This remained the determining element of all contributions of the early St Petersburg Mathematical School on the subject of approximation theory.

[Cheb59] was the only work by Chebyshev devoted to a general problem of uniform approximation theory. But it was followed by a series of more than 40 publications in which he dealt with the solution of special uniform approximation problems, mainly from the theory of mechanisms.

Another major part of his work was devoted to least squares approximation theory with respect to a positive weight function θ. In his contribution 'About continuous fractions' [Cheb55/2] he proved that (as we say now) the orthogonal projection of a function is its best approximation in the space $L_2(\theta)$. In a number of subsequent papers he discussed this fact for certain orthonormal systems and defined general Fourier expansions.

He merged the theoretical results in the publication 'On functions deviating the least possible from zero' [Cheb73], in which he determined the monotone polynomial of given degree and the first coefficient which deviates as little as possible from zero. This was the first contribution to what we know now as shape preserving approximation theory.

All of Chebyshev's work was aimed at delivering useful solutions to practical problems. The above-mentioned contributions all arose from practical problems, e. g., from the theory of mechanisms or ballistics. A small part of his work was devoted to problems from geodesy, cartography or other subjects. This ambition pervaded all of Chebyshevs work: to him practice was the 'leader' of mathematics, and he has always demanded from mathematics that it should be applicable to practical problems. Apparently he did not view this as being in contradiction to his early theoretical work, which had been devoted mainly to number theory. So we can speculate about his concept of application.

On the other hand, he clearly disassociated himself from contemporary attempts, mainly on the part of French and German mathematicians, to define the basic concepts of mathematics clearly and without contradictions. Chebyshev qualified the discussion about infinitely small quantities as 'philosophizing'. It follows that his methods, without exceptions, were of an algebraic nature; he did not mention limits except where absolutely necessary. A characteristic feature of his work is the fact that, if convergence was intuitively possible, to him it was self-evident. Thus, he often omitted to point out that an argument was valid if a function converged (uniformly or pointwise).

Besides his scientific achievements which also extended into probability theory, Chebyshev distinguished himself as a founder of a mathematical school. The first generation of the generally so-called Saint Petersburg Mathematical School only consisted of mathematicians who began their studies during Chebyshev's lifetime and were completely influenced by his work, but even more by his opinion about what mathematics should be. In Aleksandr Nikolaevich Korkin, the eldest of the schools' members, we have a truly orthodox representative of the algebraic orientation. For example he referred to modern analytic methods of treating partial differential equations as 'decadency' because they did not explicitly solve explicit equations. Other members also disassociated themselves from new mathematical directions, most notably Aleksandr Mikhajlovich Lyapunov, who sweepingly disqualified Riemannian function theory as 'pseudogeometrical'.

However, this radical position was not typical of all students of Chebyshev. Egor Ivanovich Zolotarev showed an interest in basic mathematical questions, both in his written work, where we see his deep knowledge of function theory, and in his lectures, where he endeavoured to define concepts like the continuous function as early as in the 1870s. Julian Sochocki's work was even exclusively devoted to the theory of functions in the manner of Cauchy and later these results were used by others (e.g., by Posse) to prove Chebyshev's results in a new way.

Nevertheless the Saint Petersburg Mathematical School was characterized by the orientation towards solving concretely posed problems to get an explicit formula or at least a good algorithm which is suitable for practical purposes.

It is not surprising, then, that the contributions of the members of the Saint Petersburg Mathematical School were predominantly of a classical nature and employed almost exclusively algebraic methods.

This is particularly true for the schools' work on approximation theory. Outstanding examples include the papers by Zolotarev and the brothers Andrej Andreevich Markov, Sr. and Vladimir Andreevich Markov, which were devoted to special problems from the field of uniform approximation theory.

Zolotarev investigated the problem of determining the polynomial of given degree which deviates as little as possible from zero while having its two highest coefficients fixed. Thus he directly followed the objective set by Chebyshev himself. Andrej Markov's most important contribution on this subject was the determination of a polynomial least deviating from zero with respect to a special linear condition of its coefficients. Vladimir Markov generalized this problem and solved his brother's problem for any linear side condition.

The above-mentioned three contributions were all distinguished by the fact that they presented a complete theory of their problems. This is most distinctly illustrated by the work of Vladimir Markov, who proved a special alternation theorem in this context, as well as another theorem which in fact can be called a first version of the Kolmogorov criterion of 1948. As the most important result of Vladimir Markov's paper we today acknowledge the inequality estimating the norm of the kth derivative of a polynomial by the norm of the polynomial itself. Later Sergej Natanovich Bernstein used this result to prove one of his quantitative theorems. However, consistent with the nature of the task, these investigations remained basically algebraic.

The last contribution to early uniform approximation theory coming from Saint Petersburg were Andrej Markov's 1906 lectures 'On Functions Deviating the Least Possible from Zero', [MarA06] where he summarized and clearly surveyed the respective results of Saint Petersburg mathematicians. For the first time a Petersburg mathematician presented the uniform approximation problem as a problem of approximating a continuous function by means of polynomials and proved a more general alternation theorem. Conspicuously, however, he soon returned to problems of the Chebyshev type. It is somewhat amazing that he never referred to any of the results achieved by Western European mathematicians in this context. Even the basic Weierstrassian theorem of 1885, which states the arbitrarily good approximability of any continuous function by polynomials, was not cited in these lectures.

It thus emerges with particular clarity that, because of its narrow setting of problems and its rejection of analytical methods, the uniform approximation theory of functions as developed by the Saint Petersburg Mathematical School ended in an impasse at the beginning of the twentieth century.

Outside Russia, approximation theory had other roots. A more theoretical approach had been preferred abroad because of the strong interest in basic questions of mathematics generated since the end of the 18th century by the problem of the 'oscillating string'.

The clarification of what is most likely the most important concept in modern analysis—the continuous function—generated intense interest in the consequences resulting from Weierstrass' approximation theorem. The latter had defined the aim; one now tried to make use of it, i.e., to find explicit and simple sequences of algebraic or trigonometric polynomials which converge to a given polynomial. Secondly it had to be determined how fast these sequences could converge, how fast the approximation error decreases. Such were the objectives of the long series of alternative proofs which emerged early after the original work of Weierstrass.

Natural candidates for such polynomial sequences were the Lagrange interpolation polynomials and the Fourier series.

It is surprising that a first positive result was found for the Fourier series, although the existence of continuous functions with a divergent Fourier series had been known since 1876. Lipót Fejér showed in 1900 that every function could be approximated by a version of its Fourier series, where the sum was taken from certain mean values of the classical Fourier summands.

For the case of interpolating polynomials the question likewise seemed to be negatively answered by the results of Runge [Run04] and Faber [Fab14]. However, it was again Fejér who showed that for every continuous function the sequence of the 'Hermite–Fejér Interpolants' (as we now call them) converges to the function itself.

Chebyshev's results became more commonly known in Western Europe only after the first edition of his collected works in 1899. With the introduction of analytical methods his findings could be theoretically expanded by the work of Hilbert's pupil Paul Kirchberger in 1902 [Kir02], and Émile Borel in 1905 [Bor05].

We call Dunham Jackson the founder of the quantitative approximation theory which is designed, inter alia, to determine the degree of the approximation error subject to specific requirements on the approximating function. Jackson proved a series of direct theorems in his doctoral dissertation of 1911 [Jack11]. Actually it was Hilbert's pupil Sergej Natanovich Bernstein who, a little earlier, had proved theorems of this kind—he is today considered the author of the inverse theorems which laid the foundation of the constructive function theory that characterizes functions by the order of their approximation error.

The roots of the constructive function theory lay in a very special-looking problem to which Lebesgue's proof of Weierstrass' theorem had already attracted attention: the approximation of the function $|x|$. In 1898 Lebesgue proved Weierstrass' theorem by initially approximating a continuous function by polygon lines and subsequently proving that a polygon line can be arbitrarily well approximated by polynomials [Leb98].

In the years that followed, the question of how fast $|x|$ can be approximated and the answer given by Bernstein gave rise to investigations which connected the approaches of Chebyshev and Weierstrass, that is, algebraic and analytic ideas. Especially the way he used the results of Vladimir Markov led to the insight that the degree of approximation reveals certain properties of a function. Thus, approximation theory, born from practical mechanics, helped to solve important basic mathematical problems.

Bernstein's results completed the frame of modern approximation theory, as first described in de la Vallée-Poussin's lectures from 1919, within which the theory has remained to this day [Val19].

The following key theses are first presented and substantiated in the present book:

1. The chief aim of the activities of the Saint Petersburg Mathematical School around P. L. Chebyshev on the subject of approximation was to

determine the polynomial of nth degree which deviates as little as possible from zero while having an arbitrary but fixed number of given leading coefficients.

2. This aim prevented the application of modern analytical methods to approximation theory during the early period in Saint Petersburg.

3. A merger of Weierstrass' and Chebshev's approaches was first achieved by Bernstein. Thus, we see that the Göttingen School around David Hilbert and Felix Klein had a decisive influence on the early development of approximation theory.

The present book is structured as follows:

1. In the first chapter we discuss the two works that may be considered forerunners of uniform approximation theory: Euler's cartographic investigations and Laplace's geodetic problem.

2. The second chapter is dedicated to the work of P. L. Chebyshev: His most important contributions to the uniform approximation problem are analysed and arranged in historical context. In addition, Chebyshev's philosophy of mathematics is discussed.

3. The work of Chebyshev's students founding the Saint Petersburg Mathematical School is reviewed in chapter three. We analyse in what manner they continue Chebyshev's work and adopt his aims. It becomes clear that the ideas of the mathematicians of the Saint Petersburg Mathematical School are not perfectly homogeneous. We examine their opinion about basic principles of mathematics, especially the concept of a function.

4. The absolutely different development of approximation theory in Western Europe is summarized in the fourth chapter. Starting from the problems connected with Weierstrass' approximation theorem, I have focused on questions of approximative representations of functions and their (uniform) convergence. The role of the undoubted centre of mathematics of that time, the Göttingen School around David Hilbert and Felix Klein, in the development of approximation theory is outlined on the basis of material from several archives.

5. The fifth chapter addresses the thesis that the framework of the foundation of modern approximation theory was shaped by the contributions of Bernstein. The content of what he called 'Constructive Function Theory' is described. We shall see that he achieved the link between Chebyshev's and Weierstrass' approaches.

Contents

1

Forerunners

1.1 Euler's Analysis of Delisle's Map

Cartography arose in the beginning of the 18th century in Russia with the first map covering the whole Russian empire; it was made by I. K. Kirillov in 1734 and held the scale 1:11,7 Mill. In 1745 it was followed by the "Mappa Generalis Totius Imperii Russici" (1:8,9 Mill.) developed by the St Petersburg Academy of Sciences under the main supervision of the astronomer Joseph Nicolas Delisle (1688-1768), and with co-operation of Leonhard Euler[1] (see [KayoJ]).

1.1.1 The Delislian Projection

S. E. Fel' [Fel60, p. 187] describes the construction of the maps of this atlas: "The maps of this atlas are drawn in the cone projection which preserves distances and is attributed to J. Delisle [...] The main scale is preserved in the two cutting parallels and all meridians. The map of Russia covers the region between $40°$ and $70°$ of northern latitude, thus the middle parallel lies at $55°$, and the cutting parallels are chosen at $47°30'$ and $62°30'$. So they have equal distance to the middle and the outer parallels. The meridians are divided with preservation of distances."

This kind of projection is often used when it is necessary to project a big map like the Russian empire.

Stereographic projection often used to map polar regions have the following advantages and disadvantages:

---

[1] Leonhard Euler (*Basel 1707, †St Petersburg 1783), 1720–1724 studies of mathematics and physics at Basel university, 1727 move to St Petersburg, 1730 professorship of physics at the Academy of Sciences, 1733 professorship of mathematics as successor of Daniel Bernoulli, 1735 co-operation with the department of geography, 1741 move to Berlin, 1744 director of the department of mathematics ("mathematische Klasse") of the Berlin Academy of Sciences, 1766 again professorship at the St Petersburg Academy of Sciences.

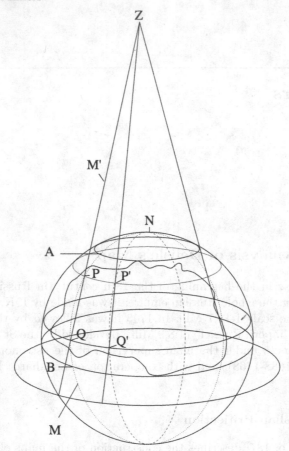

Figure 1.1. *Scheme of the Delislian cone projection with the cutting parallels PP'* and QQ'

1. Advantages:
    a) Parallels and meridians intersect perpendicularly
    b) It gives locally good approximation
2. Disadvantages:
    a) The latitudes are not equally long because the scale grows from the center to the border of the map
    b) In the case of an equatorial projection (or another non-polar projection) the meridians curve to the borders of the map. Thus taking details from such a map does not make much sense

Because of the latter disadvantage one would have to choose polar projections for an overall map of Russia. But because of the growing scale one would get a global inaccuracy of the map.

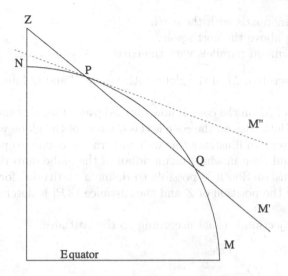

Figure 1.2. *Distance preserving division of meridians*

The map of the entire Russian Empire drawn by J. M. Hasius[2] is a polar projection. Euler had it in mind as he compared the different projections.

On the other hand, the Delislian projection meets the following claims:

- All meridians are represented by straight lines
- All parallels have the same size
- Meridians and parallels intersect perpendicularly

In 1777 Euler analysed the local and global accuracy of the Delislian conic projection in the contribution *De proiectione geographica De Lisliana in mappa generali imperii Russici usitata* [Eul77], where he tried to approximate the proportion of longitudes and latitudes of the map to the real proportion of the terrestrial globe.

The Delislian conic projection usually has cutting parallels with equal distance to the center and the borders of the map, as described above. Then the error of the proportions within the section between the cutting parallels is smaller then between the borders and the cutting parallels (see [Fel60, p. 187]).

1.1.2 Euler's Method

Now we want to get into a more exact analysis of Delisle's conic projection.

Consider a cone with the following properties (see figure 1.1):

---

[2] In 1739 Johann Matthias Hasius (1684–1742) published in Nuremberg the "Imperii Russici et Tatariae universae tam majoris et Asiaticam quam minoris et Europaeae tabula" (cited after [Eul75, p. 583, entry 195]).

1. It has a common axis with the earth.
2. Its top Z lies above the north pole.
3. It has two common parallels with the earth.

Now take a meridian M of the globe with points P and Q intersecting the cone.

The map M' of M on the cone is now divided preserving distances, that is, every distance of latitudes on the cone is the same as of the globe with respect to a constant factor. To illustrate this we can turn the cone into position M'' (see figure 1.2) and then unwind the meridian of the globe onto the cone.[3]

Using this construction it is possible to define a "latitude" for the top of the cone Z, since the position of Z and the distance $|\overline{ZP}|$ is determined by P and Q.

Thus, the projection is exact according to the latitudes.

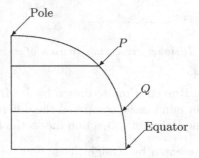

Figure 1.3. *The proportion between longitudes and latitudes from the equator to the pole is the cosine of the latitude to 1.*

To analyse the error of the longitudes on the cone, which is especially obvious regarding the pole, we will compute the length of one degree of longitude on the map.

On the surface of the globe the proportion between longitudes is the cosine of the latitude to 1. (see figure 1.3). One degree of longitude on the parallel PP' has therefore the length $\delta \cos p$, if δ is the length of one degree of latitude on the surface of the globe. On the map the length of one degree of longitude on the parallel PP' is $\omega|\overline{ZP}|$, if ω is the angle corresponding with one degree of longitude on the map (see figure 1.4).

Euler constructed a map where the maximal error of longitudes was minimized by a suitable selection of intersecting parallels P and Q.

---

[3] Then the cone's meridian is stretched by a factor which is equal to the quotient of the geodetic and the Euclidean distance between P and Q.

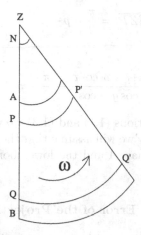

Figure 1.4. *Construction of one degree of longitude*

1.1.3 Determining the Intersecting Parallels P and Q

To construct with the above defined least maximal error we have to derive
assumptions for the positions of the points P and Q.

With this we define p and q as the latitudes of P and Q on the globe.
The length of the distance $\overline{ZP}$ and thus the position of the conic pole Z is
then

$$\frac{|\widehat{QQ'}| - |\widehat{PP'}|}{|\overline{PQ}|} = \frac{|\widehat{PP'}|}{|\overline{ZP}|},$$

that is,

$$|\overline{ZP}| = \frac{|q - p|\cos p}{\cos q - \cos p}. \qquad (1.1)$$

Now we determine the angle ω which corresponds to a degree of longitude
on the map. It is

$$\omega = \frac{|\widehat{PP'}|}{|\overline{ZP}|}$$

and with (1.1)

$$\omega = \frac{\delta(\cos q - \cos p)}{|q - p|}, \qquad (1.2)$$

δ being the length of a latitude of the globe.
Let z be the distance between Z and the Earth's pole on the globe.

The assumption that our projection preserves the latitude allows us to
compute

$$|\overline{ZP}| = \frac{\pi}{2} - p + z.$$

Using (1.1) we get

$$z = \frac{|q - p| \cos p}{\cos q - \cos p} - \frac{\pi}{2} + p. \tag{1.3}$$

With the help of the equations (1.1) and (1.3) we will determine the positions of P and Q. Additionally we will assume that the errors of the projection at the upper border of the map A and the lower border B will be equal in value.

1.1.4 Minimization of the Error of the Projection

Firstly we will compute the error in the border parallels A and B. We set a and b as their latitudes.

Their distances from the Earth's pole can be computed as above: $\frac{\pi}{2} - a$ and $\frac{\pi}{2} - b$, respectively, similarly we have $|\overline{ZA}| = \frac{\pi}{2} - a + z$ and $|\overline{ZB}| = \frac{\pi}{2} - b + z$. To get the arc length of a degree of longitude δ_a and δ_b in A and B respectively, we must multiply these values with ω.

Hence we have:

$$\delta_a = \omega |\overline{ZA}|$$
$$= \omega(\frac{\pi}{2} - a + z) \tag{1.4}$$
$$= \frac{\delta(\cos q - \cos p)(\frac{\pi}{2} - a + z)}{|q - p|},$$

$$\delta_b = \omega |\overline{ZB}|$$
$$= \omega(\frac{\pi}{2} - b + z) \tag{1.5}$$
$$= \frac{\delta(\cos q - \cos p)(\frac{\pi}{2} - b + z)}{|q - p|}.$$

But the exact values would be (see fig. 1.3) $\delta \cos a$ and $\delta \cos b$.

We remember that we wanted to determine P and Q under the assumption that the errors reach their maximal value in both border parallels A and B.

Therefore we can set

$$\omega(\frac{\pi}{2} - a + z) - \delta \cos a = \omega(\frac{\pi}{2} - b + z) - \delta \cos b$$

or

$$\omega = \delta \frac{\cos a - \cos b}{b - a}. \tag{1.6}$$

Thus we have a representation of ω by the known parameters a and b.

The next is to determine another point $X \in \overline{AB}$, where the error of the projection reaches its maximal value, too. With x defining its corresponding latitude we can define the error function

$$\varepsilon(x) : [b, a] \longrightarrow \mathbb{R} \text{ with}$$

$$\varepsilon(x) := \omega(\frac{\pi}{2} - x + z) - \delta \cos x. \tag{1.7}$$

We have then:

$$\varepsilon'(x) = \delta \sin x - \omega$$
$$\varepsilon''(x) = \delta \cos x > 0 \text{ in } [b, a],$$

so the error function has a local minimum at $\bar{x} := \arcsin \frac{\omega}{\delta}$.[4]

To determine P and Q with the above-mentioned assumptions we set

$$|\varepsilon(\bar{x})| = |\varepsilon(a)| = |\varepsilon(b)|.$$

Thus,

$$\varepsilon(\bar{x}) = -\varepsilon(a), \text{ hence}$$
$$(\frac{\pi}{2} - \bar{x} + z)\omega - \delta \cos \bar{x} = (a - \frac{\pi}{2} - z)\omega + \delta \cos a \text{ and finally}$$
$$z = \frac{1}{2} \left(\frac{\delta(\cos a + \cos \bar{x})}{\omega} + a + \bar{x} - \pi \right). \tag{1.8}$$

Given a and b we now can construct the map by firstly determining the top of the cone Z with latitude z, then dividing the meridian in equal sections of latitude δ. The basic meridian has so been drawn. The parallels are now constructed using ω as the approximation of the angle for one degree of longitude. On the parallel with latitude y one degree of longitude is approximated by $(\frac{\pi}{2} - y + z)\omega$.

With this *alternation property* Euler constructed a best possible approximation for a map of the Russian Empire which satisfies the above-mentioned side-conditions. In the next subsection we show an easy proof for this fact.

Euler himself concluded his contribution remarking that an advantage of this is the fact that the distances between any two different points is sufficiently well approximated. Therefore he would prefer this map as a suitable map for all of Russia (see [Eul77, p. 25]).

---

[4] This is the only local minimum within the interval $[b, a]$, since $a - b < \frac{\pi}{2}$. From $\varepsilon(p) = \varepsilon(q) = 0$ and $p, q \in [b, a]$ follows: $\varepsilon(\bar{x}) \leq 0 \leq \varepsilon(a) = \varepsilon(b)$. This estimate and the continuity of ε guarantee that $\bar{x} \in (b, a]$.

1.1.5 Discussion

Euler investigated the problem

$$\min_{p\in\mathbb{P}_1}\ \max_{x\in[a,b]}\ |p(x)-\delta\cos x|, \tag{1.9}$$

where δ is the length of one degree of latitude of the globe.

The alternation theorem implies that the error function

$$\varepsilon := p - \delta\cos$$

has at least three deviation points, where ε reaches its maximal value with alternating sign. For the borders of the interval $[a,b]$ there holds the side-condition

$$0 \le a < b \le \frac{\pi}{2},$$

and therefore ε has only one local minimum $\bar{x}$ in $[a,b]$, so the points a,b and $\bar{x}$ are the only deviation points.

Euler computed ε, got for $x \in [a,b]$ the formula

$$\varepsilon(x) = \omega(\frac{\pi}{2} + z) - \omega x - \delta\cos x \tag{1.10}$$

and computed the local minimum

$$\bar{x} = \arcsin(\frac{\omega}{\delta}).$$

With the help of the side-condition

$$\varepsilon(a) \quad = \quad -\varepsilon(\bar{x}) \quad = \quad \varepsilon(b) \tag{1.11}$$

he determined ω and z. As we now know, the alternation theorem shows that exactly this condition characterizes the solution of the so-defined minimax-problem.

Euler computed this minimal maximal error and got a value of about 0.0098 expressed as a part of a longitude. At the lower border of the map we have a deviation of 835 metres, at the upper border 372 metres. Euler himself had to round off and got a slightly different value of about one Russian Versta (about 1,067 metres).[5] Euler determined the intersecting parallels as 65°4'11" and 43°59'20".

Even nowadays the Delislian projection is used for entire maps of Russia. Salistshev [Sal67, p. 36] writes: "The usual projection is the conic projection which preserves distances along the meridians [...] with the intersecting parallels 47° and 62° of northern latitude which are chosen by the condition that the squares of the deviation of longitudes within the mainland of the USSR should be as little as possible." With these words he refers to to figure 1.5 taken from the second volume of [Sal67, Abb. 30].

---

[5] [Eul77, p. 295]: „et quoniam iste error in partibus unius gradus Meridiani exprimitur, 15 milliaria tali gradui tribuendo, iste error valebit 0,14190, hoc est circiter septimam partem unius milliaris, sive unam Verstam Ruthenicam".

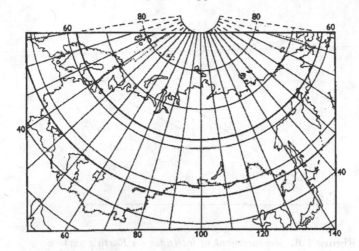

Figure 1.5. *Map of the entire USSR, drawn with a conic projection*

1.2 Laplace's Approximation of Earth's Surface

In his main work, the "celestial mechanics,"[6] Laplace[7] discussed many questions dealing with shape and orbits of celestial objects. He theoretically showed, e. g., that planets have to have the shape of an ellipsoid, but was not able to confirm this empirically.

Therefore he wanted to determine a best ellipsoid for each planet, that is, to approximate the earth with an ellipsoid by minimizing the error between the meridians of the ellipsoid and the measured meridians of the earth.

1.2.1 A Formula to Compute a Part of the Arc of an Ellipse

To determine the perimeter of the earth at a fixed meridian it is necessary to measure the length of an arc between one point with latitude $\varphi - \varepsilon_0$ to another point with latitude $\varphi + \varepsilon_0$ (see fig. 1.6).

In the following we want to compute how long this arc should be, if it were a part of an ellipse.

---

[6] „Traité de la mécanique céleste", [Lap43]. Here the basis source is §39, pp. 126–154.

[7] Pierre Simon Marquis de Laplace (*1749 Beaumont-en-Auge, †1827 Paris), 1765–1768 student at the Jesuit-college of Cáën, 1768 move to Paris, 1771–1794 teacher at the military academy of Paris, from 1773 member of the Pais academy of sciences, from 1794 professor of mathematics at the recently founded École Polytechnique, chairman of the French state commission for weights and measures, 1799 for six weeks Napoleon I's minister of the interior

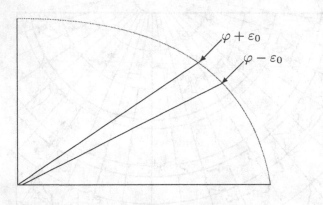

Figure 1.6. *Measurement of latitudes on Earth's surface*

1.2.1.1 The arc length of a part of an ellipse

An ellipse with the half axes a and b satisfies the parametric form

$$x(\varphi) = a \cos \varphi$$
$$y(\varphi) = b \sin \varphi. \qquad (1.12)$$

Setting $u(\varphi) := (x(\varphi), y(\varphi))$, we get the following formula for the length of an arc from the angle $\varphi - \varepsilon_0$ to $\varphi + \varepsilon_0$:

$$L(\varphi) = \int\limits_{\varphi - \varepsilon_0}^{\varphi + \varepsilon_0} \|\dot{u}(t)\| \, dt. \qquad (1.13)$$

With the numerical eccentricity of the ellipse $\varepsilon := \frac{\sqrt{a^2 - b^2}}{a}$, we have after some transformations

$$L(\varphi) = a \int\limits_{\frac{\pi}{2} - \varphi - \varepsilon_0}^{\frac{\pi}{2} - \varphi + \varepsilon_0} \sqrt{1 - \varepsilon^2 \sin^2 t} \, dt = E(\frac{\pi}{2} - \varphi - \varepsilon_0, \varepsilon) - E(\frac{\pi}{2} - \varphi + \varepsilon_0, \varepsilon), \quad (1.14)$$

where $E(\psi, k)$ is the elliptic integral of second kind. There holds the series expansion (comp., e. g. [GrHo75])

$$E(\psi, k) = \sum_{\nu=0}^{\infty} \binom{\frac{1}{2}}{\nu} (-k^2)^\nu \int\limits_0^\psi \sin^{2\nu} u \, du. \qquad (1.15)$$

Taking the first two sums of the series (1.15) we get this approximation for $L(\varphi)$:

$$L(\varphi) = a\left[\frac{\pi}{2} - \varphi + \varepsilon_0 - \varepsilon^2\left(\frac{1}{2}\left(\frac{\pi}{2} - \varphi + \varepsilon_0\right) - \frac{1}{4}\sin(\pi - 2\varphi + 2\varepsilon_0)\right)\right]$$
$$- a\left[\frac{\pi}{2} - \varphi - \varepsilon_0 - \varepsilon^2\left(\frac{1}{2}\left(\frac{\pi}{2} - \varphi - \varepsilon_0\right) - \frac{1}{4}\sin(\pi - 2\varphi - 2\varepsilon_0)\right)\right],$$

hence

$$L(\varphi) = 2a\varepsilon_0 - 2a\varepsilon^2\varepsilon_0 + \frac{a}{2}\sin 2\varepsilon_0 \cos(\pi - 2\varphi).$$

So we have

$$L(\varphi) = 2a\varepsilon_0 - 2a\varepsilon^2\varepsilon_0 + \frac{a}{2}\sin 2\varepsilon_0 + a\sin 2\varepsilon_0 \sin^2\varphi. \qquad (1.16)$$

The length of an arc of an ellipse between the angles $\varphi - \varepsilon_0$ and $\varphi + \varepsilon_0$ can be approximated by a quadratic equation of the form $z + y\sin^2\varphi$, where z is the equatorial arclength ($\varphi = 0$).

Laplace used this approximation for his computations.

1.2.2 A Characteristic System of Equations

Let n be the number of measurements on the Earth's surface with all arcs representing the length of one degree of the corresponding meridian ($\varepsilon_0 = 0,5^o$). Let a_i, $i = 1,\ldots,n$ be the measured values with the square sinus p_i, of the corresponding latitudes. We set the order $p_{i+1} > p_i \ \forall i = 1,\ldots,n-1$. Now we have the unknown parameters z as the arclength of the ellipse between $-0,5^o$ and $+0,5^o$, the increment of the measurement y and the corresponding errors x_i, $i = 1,\ldots,n$.

We showed in the paragraph before that there holds for each degree a_i of a meridian of the ellipse

$$a_i = z + p_i y \qquad (1.17)$$

(with $y = a\sin 1^o$ and $p_i = \sin^2\varphi_i$). So the above chosen parametres satisfy the following linear equation system:

$$\text{(A)} \quad \begin{array}{l} a_1 - z - p_1 y = x_1 \\ a_2 - z - p_2 y = x_2 \\ \cdots\cdots\cdots\cdots\cdots \\ a_n - z - p_n y = x_n \end{array}.$$

So we have n equations for the $n + 2$ unknown quantities z, y and $x_1,\ldots,x_n$.[8]

---

[8] The equation system (A) has the form

Firstly we eliminate the parameters z and y and get $n - 2$ equations with $x_1, \ldots, x_n$ as the only unknown.[9]

Then we want to determine the equation systems, where $\max_{i=1,\ldots,n} |x_i|$ reaches its minimal value.

The best approximating ellipse is chosen only by considering the maximal error of the x_i and deriving necessary conditions for this minimax problem. The quantities z and y are then computed using these conditions. Let in the following $\hat{x} := \max_i |x_i|$.

1.2.3 A First Solving Algorithm

To determine the solution of (A) with the least maximal error $\hat{x}$ Laplace first considered cases with a low number of measurements.[10]

The cases $n = 3, 4, 5$ were discussed separately.

In the case that we made only three observations we can say that $\hat{x}$ reaches the smallest value, if all errors have the same value (neglecting their signs).

It is indeed easy to show that this property is necessary for the minimality of $\hat{x}$.[11]

Without an alternation property this condition is not sufficient, as we will see later on. Laplace did not mention this, but his formulation suggested that the equivalence holds. This remark also holds for the following considerations.

$$\mathfrak{A}\mathbf{x} = \mathbf{a}, \quad \text{with} \quad \mathfrak{A} = \begin{pmatrix} 1 & p_1 & 1 & 0 & \cdots & 0 \\ 1 & p_2 & 0 & 1 & \cdots & 0 \\ \vdots & \vdots & \vdots & \vdots & \ddots & \vdots \\ 1 & p_n & 0 & 0 & \cdots & 1 \end{pmatrix}, \quad \mathbf{x} = (z, y, x_1, \ldots, x_n)$$

$$\text{and} \quad \mathbf{a} = (a_1, \ldots, a_n).$$

We easily see that the lines of the matrix $\mathfrak{A}$ are linearly independent. So the solution space has dimension $n + 2 - rg(\mathfrak{A}) = 2$, hence two variables can be chosen freely.

---

[9] This is a remarkable step, for we want to compute the quantities z and y, which alone determine the ellipse.

[10] Because an ellipse is determined by at least three measurements, we assume that there holds $n \geq 3$. Laplace did not mention this explicitly, but seemed to have taken it for granted. In all considerations the errors are enumerated at least to three. (See [Lap43], p. 127).

[11] After eliminating the parameters z and y in the first two equations we get the only equation

$$a = mx_1' + nx_2' + px_3',$$

where a, m, n and p can be assumed positive with a suitable choice of the sign of $x_i' = \pm x_i$. If we assumed that there would be a solution with $\hat{x} < \frac{a}{m+n+p}$, we would get the contradiction $a = mx_1' + nx_2' + px_3' < (m+n+p)\hat{x} < a$. So we have anyway $\hat{x} \geq \frac{a}{m+n+p}$. Indeed $x_1' = x_2' = x_3' = \frac{a}{m+n+p}$ satisfies the above-mentioned condition.

If we have only two equations determining the errors, the solution with the smallest maximal error is the one, where three errors have the same value and the fourth has a lower value than they (neglecting the signs).

Laplace proved this by tracing this case back to the first.

If we assume that x_1 is the error reaching another absolute value than the other ones, we will eliminate it and get for the other three errors the determining equation

$$a = mx_2 + nx_3 + px_4,$$

where a and all other coefficients can be chosen positively.[12]

The smallest value of the maximal error of this equation is again $a/(m+n+p)$. Now we compute the value of x_1 and check, whether its value is not larger than the absolute value of x_1, x_2 and x_3. If not, then we have to eliminate x_2 and repeat this, until the minimal error will be found.

If we have only three equations determining the errors, the solution with the smallest maximal error is the one, where three errors have the same value and the other two have lower values (neglecting the signs).

The arguments are analogous, we eliminate two errors and get a determining equation for the other three errors which has the same type as above.

This algorithm can be used for any numbers of observations.[13]

So Laplace could prove that the condition that *the three largest errors of a solution of the equation system (A) have the same absolute value is necessary for the solution with the smallest maximal error.*

He proposed an algorithm to determine the ellipse of best approximation in the above sense. Set equation system (A) and choose all triples of the errors x_i. One after the other assume that they are the maximal errors (neglecting their signs) and check the effects on the other errors. If all other errors have smaller absolute values, then the ellipse is the sought one. Because of the enormous complexity of this algorithm[14] Laplace dismissed it and proposed a more simple one.

1.2.4 A Second Algorithm and a Necessary Alternation Condition

The largest errors of equation system (A) have even to satisfy an alternation condition. Laplace showed this proving that such a condition is necessary for a solution $(z, y, x_1, \ldots, x_n)$, where the x_i represent a system of errors with smallest maximal error.

---

[12] Compare the former consideration with quantities x_i' mit $x_i = \pm x_i$.

[13] Laplace wrote (in [Lap43], p. 128): „Il est facile d'étendre cette méthode, au cas où l'on auroit quatre ou un plus grand nombre d'équation de condition, entre les erreurs $x^{(1)}$, $x^{(2)}$, $x^{(3)}$, &c".

[14] Since we also have to check all eight combinations of each triple, we must consider $8n^3$ equation systems in the worst case.

We note Laplace's considerations as lemmata to make it more clear. Laplace himself points out the main arguments without a formal proof.

For the following steps we call $\mathbf{x} := (z, y, x_1, \ldots, x_n)$ a solution of (A) with the smallest maximal error.

Lemma 1.1 *Let x_i be an error of maximal (absolute) value of $\mathbf{x}$. Then there is another error $x_{i'}$ ($i' \neq i$) which has the same absolute value and opposite sign.*

If we had only one maximal error x_i or any number of errors with equal value and equal sign we could diminish the value of these x_i in the corresponding equation

$$a_i - z - p_i y = x_i,$$

and they would remain a maximum.[15]

With the following lemma we find at least a third maximal error:

Lemma 1.2 *Between the values x_i, $i = 1, \ldots, n$ there are at least three values $\pm\hat{x}$. Their signs are not equal.*

Here we put together Laplace's proposition ,,il doit exister une troisième erreur $x^{(i'')}$ égale, abstraction faite du signe, à $x^{(i)}$." ([Lap43], p. 129) with the result of the last lemma.

We assume that there are only two maximal errors x_i and $x_{i'}$ and look at the equation which we get after subtracting the equations corresponding to x_i and $x_{i'}$ from each other:

$$a_{i'} - a_i - \{p_{i'} - p_i\}y = x_{i'} - x_i.$$

The right side of the equation is equal to $2 \pm \hat{x}$. This sum can be diminished varying y.[16]

---

[15] A formal proof could be the following:

1.) If we had only one x_j with $|x_j| = \hat{x}$, then we could vary z so that

$$\max_i |x_i| < \hat{x} :$$

Replace (A) with the following equation system

$$(A') \quad \begin{aligned} a_1 - z - \tfrac{\varepsilon}{2} - p_1 y &= x_1' \\ a_2 - z - \tfrac{\varepsilon}{2} - p_2 y &= x_2' \\ &\cdots\cdots\cdots\cdots\cdots \\ a_n - z - \tfrac{\varepsilon}{2} - p_n y &= x_n' \end{aligned}$$

with $\varepsilon := \hat{x} - \max_{i \neq j} |x_i|$. Then we have for (A'): $\max_{i=1,\ldots,n} |x_i'| < \hat{x}$, so we diminished the maximal error.

2.) If more than one error have the value $\hat{x}$ and are equal in sign and all other errors have smaller absolute values, then the procedure of 1.) works equally. □

[16] Assume that $i' > i$. Then the coefficient of y is positive, since $p_{i'} > p_i$ because they are squares of sines of the first quadrant.

Now we can replace $x'_{i'}$ and x'_i with the recently computed smaller errors and set them equal in absolute value by varying z to get an equation system with smaller maximal error.

Now we know that the (at least three) maximal errors have different signs. The following shows that the signs are not ordered by chance.

Lemma 1.3 (Alternation) *There are three maximal errors x_i, $x_{i'}$ and $x_{i''}$ of the solution* **x** *of equation system (A) with the following property: If x_i and $x_{i''}$ have an equal sign, so the index i' lies between the indices i and i''.*

So we have a system of errors with alternating signs.

To prove it we assume that for all error triples having different signs we have that i' is either larger or smaller than the other two indices i and i''. We now regard the equations which we get after subtracting the equations corresponding to i' from the equations corresponding to i and i'' respectively. Then we have

$$a_i - a_{i'} - (p_i - p_{i'})\,y = x_i - x_{i'}$$
$$a_{i''} - a_{i'} - (p_{i''} - p_{i'})\,y = x_{i''} - x_{i'}.$$

Both right sides are equal to $2\,\hat{x}$ and have the same sign. But then again we can diminish the error sum and the maximal error itself varying z[17] contradicting the assumption **x** being a solution of (A) with the smallest maximal error.

In the following Laplace used this result for the explicit determination of the maximal errors.[18]

1.2.5 Determination of the Maximal Errors

1.2.5.1 Determination of the Largest Error

To find out which of the errors x_1, x_2, ... , x_n has the maximal value with respect to the sign, we consider the equation system

$$\text{(B)} \quad \begin{aligned} a_2 - a_1 - (p_2 - p_1)\,y &= x_2 - x_1 \\ a_3 - a_1 - (p_3 - p_1)\,y &= x_3 - x_1 \\ &\cdots\cdots\cdots\cdots\cdots \\ a_n - a_1 - (p_n - p_1)\,y &= x_n - x_1 \end{aligned}.$$

---

[17] After simultaneously diminishing the error sum we can vary z so that $x_{i'}$ keeps its value and that the values of x_i and $x_{i''}$ are reduced. Then $x_{i'}$ is the only maximal error which contradicts Lemma 1.

[18] Laplace in fact did not show the sufficiency of the alternation criterion. But as we will see later on in this book, it can be shown easily by using similar 'shifts' as in the present proofs.

Since all coefficients of y are positive, for large y all right sides are negative and x_1 is the largest error. We now can diminish y so that some of the right sides will vanish, that is, some of the x_i will assume a value equal to x_1. To determine these x_i consider the quotients

$$\frac{a_2 - a_1}{p_2 - p_1}; \quad \frac{a_3 - a_1}{p_3 - p_1}; \quad \ldots \ldots; \quad \frac{a_n - a_1}{p_n - p_1}.$$

We set β_1 equal to the maximal of these quotients and set r as the first of all indices relating to all quotients having value β_1.

Hence

$$\beta_1 = \max_{i>1} \frac{a_i - a_1}{p_i - p_1} = \frac{a_r - a_1}{p_r - p_1}. \tag{1.18}$$

If we now replace in (B) $y := \beta_1$, we will have $x_r = x_1$, and by further reduction of y, x_r will get larger than x_1 and grow faster than all other errors with smaller index. This directly follows from the definition of β_1.[19]

In other words:

With further reduction of y the maximal error in (B) will only grow for indices larger than r.

After the determination of x_r we similarly continue with this procedure and consider the equation system

$$a_{r+1} - a_r - (p_{r+1} - p_r)\, y = x_{r+1} - x_r$$
$$a_{r+2} - a_r - (p_{r+2} - p_r)\, y = x_{r+2} - x_r$$
$$\cdots\cdots\cdots\cdots\cdots\cdots\cdots\cdots\cdots\cdots .$$
$$a_n - a_r - (p_n - p_r)\, y = x_n - x_r$$

---

[19] To prove the last proposition we consider the equations

$$\frac{a_{r-t} - a_1}{p_{r-t} - p_1} - y = \frac{x_{r-t} - x_1}{p_{r-t} - p_1}$$
$$\frac{a_r - a_1}{p_r - p_1} - y = \frac{x_r - x_1}{p_r - p_1}, \quad 0 < t < r - 1.$$

For $y < \beta_1$ the right side of the rth equation is larger than the right side of the $r - t$th. Without losing generality we assume $x_{r-t} > x_1$ (otherwise we will have immediately $x_{r-t} \leq x_1 < x_r$!).

Then:

$$\frac{x_r - x_1}{p_r - p_1} > \frac{x_{r-t} - x_1}{p_{r-t} - p_1}, \quad \text{from which follows}$$
$$\frac{x_r - x_1}{x_{r-t} - x_1} > \frac{p_r - p_1}{p_{r-t} - p_1} \, (> 1), \quad \text{hence}$$
$$x_r - x_1 > x_{r-t} - x_1 \quad \text{and finally}$$
$$x_r > x_{r-t}.$$

Analogously we determine

$$\beta_2 := \max_{i=1,\ldots,n-r} \frac{a_{r+i} - a_r}{p_{r+i} - p_r}$$

and call r_2 the first index for which the quotients assume the value β_2. Between $y = \beta_1$ and $y = \beta_2$ the error reaches its maximal value at x_r. By further reduction of y, x_{r_2} will become maximal. If y gets smaller than β_2, for $i < r_2$ all x_i remain smaller than x_{r_2} and for larger indices some x_i will exceed it.

Proceeding again we will finally get the sets

$$\mathfrak{X} = \{x_1, x_r, x_{r_2}, \ldots, x_{r_k} = x_n\}$$
$$\mathfrak{B} = \{\infty, \beta_1, \beta_2, \ldots, \beta_k, -\infty\}$$

containing the maximal errors (respecting the signs) and the corresponding values of y. When y reaches the value β_m, x_{r_m} will be the maximal error and remains so until y reaches β_{m+1}.

1.2.5.2 Determination of the Smallest Error

Lemma 1.3 shows that between two maximal errors with equal sign there is a maximal error with opposite sign. We can use the above determined procedure to determine the smallest error (the maximal error with negative sign), too. Then we again consider equation system (B). For a fixed and arbitrarily small y all the right sides are positive and x_1 a minimum. Now we determine

$$\gamma_1 := \frac{a_{s_1} - a_1}{p_{s_1} - p_1} := \min_{i=1,\ldots,n} \frac{a_i - a_1}{p_i - p_1}.$$

As a result we will get the two sets

$$\mathfrak{X}' = \{x_1, x_{s_1}, x_{s_2}, \ldots, x_{s_l} = x_n\}$$
$$\mathfrak{C} = \{-\infty, \gamma_1, \gamma_2, \ldots, \gamma_l, \infty\}$$

containing the minimal errors (respecting the signs) and the corresponding values of y. When y reaches the value γ_m, x_{s_m} will be the minimal error and remains so until y reaches γ_{m+1}.

1.2.5.3 Determination of the Best Ellipse

We know from the previous paragraphs that the solution of the parameter y is one of the values $\beta_1, \ldots, \beta_k$ or $\gamma_1, \ldots, \gamma_l$, since between these values the maximal error has a constant value.

1.) Furthermore we record the fact that there will hold $y \in \mathfrak{B} \setminus \{\infty, -\infty\}$ in the case, if two mximum errors have positive sign and the third has negative sign. Then the positive maximal errors are subsequent elements of $\mathfrak{X}$, that is with x_{r_i} being a maximum, $x_{r_{i+1}}$ is a maximum, too.

2.) The respective statement is valid for $y \in \mathfrak{C} \setminus \{\infty, -\infty\}$. If two maximal errors have a negative sign and the third is positive, then $x_{s_{i+1}}$ will be a minimum, if x_{s_i} is one.

We can now successively choose for y the possible values from $\mathfrak{B}$ and $\mathfrak{C}$ and determine the right value for y using the following algorithm:

1. For $y = \beta_i$ take x_{r_i} and $x_{r_{i+1}}$ maximal errors. Find the value γ_k lying between β_i and β_{i+1}. If there is no such value, proceed with β_{i+1}. Otherwise, if this holds for γ_j, x_{s_j} will be the corresponding minimal error.
2. After choosing all values from $\mathfrak{B}$ proceed analogously with γ_μ. For $y = \gamma_\mu$ take x_{s_μ} and $x_{s_{\mu+1}}$ as minimal errors and find β_ν between γ_μ and $\gamma_{\mu+1}$. If there is no such value, proceed with $\gamma_{\mu+1}$. Otherwise: If β_ν is the sought value, $x_{r_{\nu+1}}$ will be the corresponding maximal error.

The size of the extremal errors can be computed subtracting the corresponding lines of the equation system (A). For it is

$$a_r - a_s - (p_r - p_s)\,y = x_r - x_s = \pm 2\,\hat{x}, \qquad (1.19)$$

if x_r is a maximal error and x_s a minimal error.

Finally we can compute z using the known values. With r and s the indices for the maximal and the minimal error we have using Lemma 2: $x_s = -x_r$, and regarding (A)

$$a_r - z - p_r\,y = x_r$$
$$a_s - z - p_s\,y = -x_r$$

we can determine

$$z = \frac{a_r + a_s}{2} - \frac{p_r + p_s}{2}\,y. \qquad (1.20)$$

With this algorithm Laplace could essentially improve the complexity of the computations. To determine $\mathfrak{B}$ and $\mathfrak{C}$ we have to compute n quotients of type (1.18). Then there follow at most $2n$ computations of type (1.19) and finally the computation of z and y. So the algorithm has order $\mathcal{O}(n)$ instead of $\mathcal{O}(n^4)$ as we had before.

1.2.6 Application to Geodesy

Laplace applied his algorithms in §41 of the „Mécanique Celéste" [Lap43] to determine the ellipsoid best approximating the surface of the earth. There he used known data from former measurements. Table 1.1 shows these data. The values are given in double toises (a toise is about 1.9946m).

After computing the quotients

$$\frac{a_j - a_1}{p_j - p_1}, \quad j = 2, \ldots, 7,$$

i	Place of Measurement	a_i	Latitude	p_i
1	Peru	25538.85	0°	0
2	Cape of Good Hope	25666.65	33.3084°	0.30156
3	Pennsylvania	25599.60	39.2000°	0.39946
4	Italy	25640.55	43.0167°	0.46541
5	France	25658.28	46.1994°	0.52093
6	Austria	25683.30	47.7833°	0.54850
7	Lapland	25832.25	66.3333°	0.83887

Table 1.1. *Using measurements of meridianial arclength in seven different places Laplace applied his formulae. The quantities a_i stand for the measured lengths of one degree of latitude, p_i are the square sines of the corresponding latitudes.*

we get the maximal value at $j = 2$. So we have $r_1 = 2$ and $\beta_1 = 423.7961$. As a result we get the sets

$$\mathfrak{X} = \{x_1, x_2, x_7, \} \text{ and}$$
$$\mathfrak{B} = \{\infty, 423.796, 308.204, -\infty\}$$

for the positive maximal errors and

$$\mathfrak{X}' = \{x_1, x_3, x_5, x_7, \} \text{ and}$$
$$\mathfrak{C} = \{-\infty, 152.080, 483.097, 547.182, -\infty\}$$

for the negative maximal errors.

We proceed comparing the errors: Choose $\beta_1 = 423.796$. Between β_1 and $\beta_2 = 308.204$ there is no value from $\mathfrak{C}$. Therefore continue with β_2 and $\beta_3 = -\infty$. Between them there is $\gamma_1 = 152.080$. So we assume x_2, x_3 and x_7 as maximal, x_3 has negative, the others positive sign.

Now we still have to compare the maximal errors with negative sign: Choose $\gamma_1 = 152.080$. Between γ_1 and $\gamma_2 = 483.097$ there lies β_1. So as maximal errors we assume x_2, x_3 and x_5. But x_3 and x_5 have equal sign, and so we have to reject this choice. The same holds for the alternative β_2 corresponding with x_7. Since the only suitable values are the outer errors x_2 and x_7, no choice for γ_i will be the correct one.

So we have already got our maximal errors, they are

$$x_2 = -x_3 = x_7 = 48.612.$$

1.2.7 A Discrete Approximation Problem

To discuss Laplace's results we have to remark that he considers a discrete approximation problem.

Assume that the function

$$b : [-\frac{\pi}{2}, \frac{\pi}{2}] \to \mathbb{R}$$

describes the latitude of one degree of the meridian at a point of the surface of the earth and let the function be continuous, so we can formulate Laplace's problem as the trigonometric approximation problem

$$\min_{y,z\in\mathbb{R}} \max_{\varphi\in[-\frac{\pi}{2},\frac{\pi}{2}]} |b(\varphi) - (z + y\sin^2\varphi)|. \tag{1.21}$$

Using discrete data (b is unknown) he determines the minimal solution for a set

$$\{\varphi_1,\ldots,\varphi_n\}$$

having observed values

$$b(\varphi_i) = a_i, \quad i = 1,\ldots,n.$$

For these angles he proves an alternation criterion that is a forerunner of de la Vallée-Poussin's theorem about the best approximation on a set of points[Val19, p. 78ff.].

If we consider only one quadrant, e. g. $[0,\frac{\pi}{2}]$), the alternation theorem generally holds for this problem, since (span $\{1,\sin^2\}$) is a Haar space in $C[0,\frac{\pi}{2}]$.

So for a known b Laplace's method can be used to determine the best approximating ellipsoid for the earth.

2

Pafnuti Lvovich Chebyshev

Up to now we have only discussed some particular cases of the uniform approximation problem, which were far from a mathematical theory. Indeed fifty years would pass, until Pafnuti Lvovich Chebyshev first posed a general problem of that theory that would then be named after him.

2.1 Chebyshev's Curriculum Vitae

Pafnuti Lvovich Chebyshev was born May 4 (16)[1] 1821 in the village of Okatovo, the district of Borovsk, province of Kaluga.[2] His father was the wealthy landowner Lev Pavlovich Chebyshev.[3]

Pafnuti Lvovich got his first education at home from his mother Agrafena Ivanovna Chebysheva (reading and writing) and his cousin Avdotya Kvintillianovna Sukhareva (French and arithmetic). Obviously his music-teacher[4]

---

[1] Until 1917 the Julian calendar was valid, which does not carry out the leap year correction every 100 years (but does every 400 years), implemented by the Gregorian calendar. So in the 19th century Russian dates deviate 12 days from Western dates and 13 days in the 20th century. Therefore we add to Russian dates the Gregorian dates in brackets.

[2] Biographical data were mostly taken from [Pru50] and [Pru76].

[3] We often find the note that Chebyshev is noble by descent. This information has to be treated carefully. Lev Pavlovich (and so his sons) had been awarded the title of a 'dvoryanin' (дворянин), but it had been the lowest title of nobility in Russia and could be awarded for outstanding merits, even in civil life. Nevertheless there are testimonies that Chebyshev's family, especially the mother Agrafena Ivanovna tried to separate itself clearly from commoners (comp. [Pru76, p. 15 ff.]).

[4] Prudnikov does not mention her name as also the cited source. The impersonal name 'music-teacher' might imply that it was not A. K. Sukhareva, since she was a close relative. So one should be cautious in comments on the role of Sukhareva in the mathematical education of young Pafnuti Lvovich.

also played an important role in Chebyshev's education, for she "accustomed his mind to exactness and analysis,[5]" as Chebyshev himself mentioned.

Possibly a physical handicap, whose reasons are yet unknown, was important for Chebyshev's adolescence and development: He limped since his childhood and walked with a stick.[6] Therefore his parents had to give up the idea to make an officer's career possible for him, although he would have followed the family's tradition.[7] His complaint excluded him from most of the usual children's games, so very soon he devoted himself to a passion which would determine his whole life: the construction of mechanisms.

In 1832 the family moved to Moscow mainly to attend to the education of their eldest sons (Pafnuti and Pavel, who would become a lawyer). The education continued at home, P. N. Pogorelski[8] was engaged as a teacher for mathematics and physics; he was considered as one of the best teachers in Moscow and, e. g., had educated the writer Ivan Sergeevich Turgenev. For the other subjects teachers with excellent reputation were invited, too.

In summer 1837 Chebyshev passed the registration examinations and in September he started the studies of mathematics at the second philosophical department of Moscow university. Among his teachers were counted N. D. Brashman,[9] N. E. Zernov[10] and D. M. Perevoshchikov.[11] No doubt that among

---

[5] Cited after the memoirs of D. I. Mendeleev [Men01]: «Покойный мой друг П. Л. Чебышев [...] вспоминая свое детство, рассказывал нам, что своим развитием объязан бывшей у него учительнице музыки, которая музыке-то его не научила, а ум ребенка приучила к точности и анализу.»

[6] His close friendship with Charles Hermite might also have been caused by this: Hermite had a similar complaint.

[7] During the Napoleonic war his father had taken part in the conquest of Paris as a cavalry ensign and had been decorated.

[8] Platon Nikolaevich Pogorelski (1800–1852), Master of mathematics, established his reputation mainly by the publication of important textbooks of mathematics, which he either translated or wrote by himself. These books were considered an outstanding set. Chebyshev was said to have this opinion about Pogorelski's text book on algebra: "It is the best book written in Russian because it is the briefest." ([Pru76, p. 33]: «это самая лучшая из всех книг на русском языке, потому что она самая краткая». Besides this it is interesting to mention that three of Chebyshev's five examinations he had to pass to study at Moscow university were based on Pogorelski's textbooks.

[9] Nikolaj Dmitrievich Brashman (1796–1866), 1825–1834 Assistant (Adjunkt – адъюнкт) for mathematics at Kazan university, since 1834 professor at Moscow university, from 1855 corresponding member of the Russian academy of sciences, works about mechanics and the theory of mechanisms, co-founder of the Moscow mathematical society and its journal Matematicheski Sbornik.

[10] Nikolaj Efimovich Zernov (1804–1862), studied at Moscow university, master thesis 1827. 1827–1834 teacher of mathematics at several schools of Moscow, 1834 Adjunkt, 1835 extraordinary professor at Moscow university. 1837 doctor thesis, 1842 ordinary professor. Papers on differential equations and higher algebra.

[11] Dmitrij Matveevich Perevoshchikov (1788–1880), studies in Kazan'. 1818–1826 lecturer, 1826–1851 professor at Moscow university, 1848–1851 rector Moscow

them Brashman had the greatest influence on Chebyshev. He instructed him in practical mechanics and probably showed him the work of Poncelet.

In 1841 Chebyshev was awarded the silver medal[12] for his work 'calculation of the roots of equations' [Cheb38]which had already been finished in 1838. In this contribution Chebyshev derived an approximating algorithm for the solution of algebraic equations of nth degree based on Newton's algorithm.

In the same year he finished his studies as 'most outstanding candidate' («отличнейшим кандидатом».[13])

In 1841 Chebyshev's financial situation drastically changed. In Russia a famine broke out, Chebyshev's parents were forced to leave the city and were not able to support their son anymore.[14] Nevertheless he decided to continue his mathematical studies and prepared the master examinations which spread over half a year. He passed the final examination in October 1843. In 1846 he defended his master thesis 'An Attempt to an Elementary Analysis of Probabilistic Theory'[Cheb45]. The biographer Prudnikov assumes that Chebyshev was directed to this mathematical branch after getting knowledge about recently published books on probabilistic theory or the revenue of the insurance industry in Russia.[15]

In 1847 Chebyshev defended his dissertation pro venia legendi "About integration with the help of logarithms" [Cheb47] from the St Petersburg university and so obtained the right to teach there as a lecturer. At that time some of Leonhard Euler's works rediscovered by P. N. Fuss were published by V. Ya. Bunyakovski,[16] who encouraged Chebyshev to engage in the studies of them.

University 1851 move to St Petersburg, 1852 adjunkt, 1855 ordinary member of the academy of sciences. Works about astronomy and mathematical physics.

[12] Prudnikov [Pru76, p. 37] critized the jury's decision as incompetent because not Chebyshev, but the student Anton Smolyak had been awarded the gold medal. Smolyak's paper disappeared, however, and therefore we should be careful in such judgements. So the authors (whose names are not mentioned) of the biographical articles of the fifth volume of Chebyshev's collected works refrained from making a comment (comp. [Chebgw2, Bd. 5, S. 193/197]).

[13] For explanations of the term 'candidate' refer to appendix B.1.

[14] Butzer and Jongmans speculate in [BuJo91] by consulting interesting information from the assets of Eugène Catalan whether Chebyshev had already been on a trip abroad in 1842. We do not find any clear indications of this in the biographies about Chebyshev.

But it is possible that his financial circumstances did not allow a journey to France or Belgium, at all. His well-known trips abroad all were supported by state authorities or were made when he had already become an ordinary professor and so had sufficient funds.

[15] See [Pru76, S. 64 f.].

[16] Viktor Yakovlevich Bunyakovski (1804–1889), doctor thesis 1825 in St Petersburg, 1827–1862 professor at the naval college (морской корпус), 1830–1846 additionally at the institute of transportation (институт путей сообщения), 1828–1830 adjunkt at the academy of sciences, from 1830 full member, from 1864 vice president, from 1858 official expert for statistics and the insurance industry. His works

So the basics for his subjects of interest were established. Already in 1848 he had submitted his work 'theory of congruences' for his doctorate, which he defended in May 1849. After one year he was elected to extraordinary professor at St Petersburg University, in 1860 he became ordinary professor. In 1872, after 25 years of lectureship, he became "merited professor" (заслуженный профессор), in 1882 he left the university and completely devoted his life to research.

Besides his lectureship at the university from 1852 to 1858 Chebyshev taught practical mechanics at the Alexander Lyceum in Tsarskoe Selo (the today Pushkin), a southern suburb of St Petersburg.

His scientific achievements, which we will discuss extensively in the following, give the reasons for his election as a junior academician (adjunkt) in 1856. Later on he became an extraordinary (1856) and in 1858 an ordinary member of the Imperial Academy of Sciences. In the same year he became an honourable member of Moscow University.

Moreover, he assumed other honourable appointments and was decorated several times: in 1856 Chebyshev became a member of the scientific committee of the ministry of national education; in 1859 it was followed by the ordinary membership of the ordnance department of the academy with the adoption of the headship of the "commission for mathematical questions according to ordnance and experiments related to the theory of shooting,"[17] the Paris academy elected him corresponding member in 1860 and full foreign member in 1874 and in 1893 he was elected honourable member of the St Petersburg Mathematical Society, recently founded in 1890.

Pafnuti Lvovich Chebyshev died November 26, 1894 in St Petersburg.

2.2 Stimuli for the Development of a Theory

The initial impulses for Chebyshev's engagement in approximation theory were set by the theory of mechanisms playing an important role at that time of arising industrialisation. Mechanisms were used in steam engines which constituted the heart of the production means, the factories.

2.2.1 Chebyshev's Trip Abroad

Already in December 1850 as a junior scientist (adjunkt) Chebyshev was appointed to a trip abroad to visit the exhibition of machines in London[18], but he still had to wait one and a half years.

---

deal with number and probabilistic theory. Next to Brashman he counted as the most important teacher of Chebyshev.

[17] [Chebgw2, p. 465]: «коммиссия по математическим артиллерийским вопросам и опытам, относящимся до теории стрельбы».

[18] This information is taken from a letter written by the dean Emil Lenz to the 'trustee' (попечитель) of the St Petersburg educational district [LeTr]

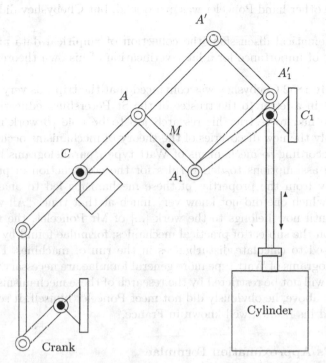

Figure 2.1. *Watt's complete parallelogram (from [ArLe55/2]), a hinge mechanism used in steam engines to tranform linear into circular movements. Filled circles mark fixed links. The bars CA, AA_1 and A_1C_1 represent the shortened form of Watt's parallelogram.*

In summer 1852 Chebyshev visited Belgium, France, England and Germany to interchange ideas about many different mathematical subjects.[19]

We have already pointed out that mechanisms very early became one of his favourite subjects, but Chebyshev had not written a paper about this, although he had even taught practical mechanics. In England and France he attempted to learn the modern scientific results on that topic.

Therefore contacts with mathematicians who had already dealt with the theory of mechanisms were of particular importance for him. In his reports he emphasized the meetings with the English mathematicians Sylvester and Cayley and the engineer Gregory. He mentioned the French mathematicians Cauchy, Liouville, Bienaymé and Hermite only in connection with his dissertation pro venia legendi 'About integration with the help of logarithms'

---

[19] The following explanations are based on Chebyshev's letters 'report about [...] the trip abroad' [Cheb52/2] and 'report [...] about the trip to England' [Cheb52/1]. Additional information is taken from the letters [LeTr], [TrCo] and [ChebTrust].

[Cheb47]. On the other hand Poncelet was mentioned, but Chebyshev did not meet him.

Besides mathematical discussions the collection of empirical data about mechanisms were of importance for a later verification of his own theoretical results.

Already in July 1852 Chebyshev was convinced that the trip was very successful. He noted in a letter to the trustee of the St Petersburg educational district[20] that he had to restrict his research job to be able to work more carefully. Probably the new discoveries of the theory of mechanisms occupied him. He wrote according to mechanisms of Watt type ('parallelograms'): He realized that the assumptions to derive rules for the construction of parallelograms directly from the properties of these mechanisms led to analytic questions about which one did not know very much at that time. "All what has been done until now belongs to the work [...] of Mr Poncelet, the well-known scientist on the subject of practical mechanics; formulae found by him are very often used to calculate disturbances in the run of machines. For a theory of parallelograms of Watt type more general fomulae are necessary, and their application will not be restricted by the research of these mechanisms.[21]"

As mentioned above, he obviously did not meet Poncelet himself. It seems that he only cited his results, well-known in France.

2.2.2 Poncelet's Approximation Formulae

For Chebyshev the fundamentals of approximation theory were established by the French mathematician Jean Victor Poncelet[22] In his work „Sur la valeur approchée linéaire et rationelle des radicaux de la forme $\sqrt{a^2 + b^2}$, $\sqrt{a^2 - b^2}$ etc."[23] [Pon35] he set the problem to approximate roots of the form $\sqrt{a^2 + b^2}$, $\sqrt{a^2 - b^2}$, and $\sqrt{a^2 + b^2 + c^2}$ *uniformly* by linear expressions. In the following we want to explain his consideration of the first case. For $\frac{a}{b} \geq k$ and $k \in \mathbb{R}_0^+$ he determined an approximation

$$\sqrt{a^2 + b^2} \sim \alpha a + \beta b \tag{2.1}$$

with real numbers α and β.

---

[20] Cited by the trustee in [TrCo].

[21] [Cheb52/2, S. 249]«Все, что сделано в этом отношении, принадлежит члену Парижской Академии Наук г. Понселэ, известному ученому в практич-еской механике; формулами, им найденными, пользуются очень много при вычислении вредных сопротивлений машин. Для теории параллел-ограмма Уатта необходимы формулы более общие, и приложение их не ограничивается исследованием этих механизмов.»

[22] Jean Victor Poncelet (*Metz 1788, †Paris 1867), studies at the École polytech-nique (one of his teachers was G. Monge), 1812 participation in Napoleon's cam-paign against Russia, arrest, 1814 return to Paris, 1815–1825 military engineer in Metz, 1825–1835 professor of mechanics at the University of Metz.

[23] Comments can be found in [Gon45] and [Gus72].

Setting $x := \frac{a}{b}$ and $r : \mathbb{R}_0^+ \ni x \mapsto \frac{\alpha x + \beta}{\sqrt{x^2 + 1}} - 1$ we can formulate Poncelet's problem as a minimax-problem:

$$\max_{x \geq k, k \geq 0} |r(x)| \quad \to \min. \tag{2.2}$$

Such a task was essentially new because it was the first formulation of a uniform approximation problem posed generally and so abstracting from a concrete problem (like Euler's or Laplace's) and enclosing a set of functions.

To solve these problems Poncelet even proved a necessary alternation condition. He established the fact that the function r is increasing for $x < \frac{\alpha}{\beta}$ and decreasing for $x > \frac{\alpha}{\beta}$. Therefore the maximum of $|r|$ is taken at the points k or $\frac{\alpha}{\beta}$ or reached by $x \to \infty$. Poncelet's calculations led to the result that all values of $|r|$ being the best approximating function must be equal. Then the alternation easily follows from the monotonicity left and right from $\frac{\alpha}{\beta}$. Poncelet proved this using geometric and analytical reflections.

2.2.3 Watt's Mechanism

At that time planar joint mechanisms based on the ideas of James Watt were the most important mechanisms to transform linear motion into circular motion. The mechanisms, however, do not execute this transformation without mistakes.

Figure 2.2 shows a shortened form of Watt's parallelogram with the connecting-rod AA_1 and the cranks CA and A_1C_1. The complete form is constructed towards the top expanding AA_1 to a parallelogram (see fig. 2.1 on page 25). Turning the crank CA round C in the direction of the crank on the opposite side and afterwards turning in the opposite direction (round C_1A_1), M follows a bow-like orbit ('Watt's curve'). If this orbit were a straight line, the transformation would be exact.

2.2.4 Watt's Curve

Simplifying the following calculations we set the length of both cranks equal, so:

$$|CA| = |A_1C_1|,$$

furthermore we assume M be the center of the connecting-rod AA_1.

This case is often regarded as standard, for instance by authors of comments about Chebyshev's work on the theory of mechanisms (comp. [ArLe55/2]), but also by Chebyshev himself (see, e. g. ,,Sur une modification du parallélogramme articulé de Watt" [Cheb61]).

We now define the length of a crank as

$$R := |CA|(= |A_1C_1|), \tag{2.3}$$

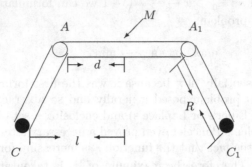

Figure 2.2. *Scheme of Watt's parallelogram in the shortened form with the movable connecting-rod AA_1*

half the length of the supporting bar

$$l := \frac{|CC_1|}{2} \tag{2.4}$$

and half the length of the connecting-rod

$$d := \frac{|AA_1|}{2}. \tag{2.5}$$

To work right the mechanism must satisfy the conditions

$$l - R < d < l + R. \tag{2.6}$$

Now define for the points A, A_1 and M coordinates

$$A = (\xi, \eta) \qquad A_1 = (\xi_1, \eta_1) \qquad M = (x, y). \tag{2.7}$$

Moving the mechanism we define as τ the angle the point A has been turned round and as τ_1 the angle the point A_1 has been turned round.

Consequently,

$$\xi = R \cos \tau - l \qquad \xi_1 = R \cos \tau_1 + l \qquad x = \frac{\xi + \xi_1}{2} \tag{2.8}$$

$$\eta = R \sin \tau \qquad \eta_1 = R \sin \tau_1 \qquad y = \frac{\eta + \eta_1}{2}. \tag{2.9}$$

With Pythagoras we have

$$(\xi - \xi_1)^2 + (\eta - \eta_1)^2 = 4d^2; \tag{2.10}$$

after substituting variables and some transformations we get

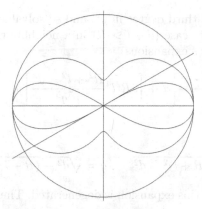

Figure 2.3. *Two examples of the graph of Watt's curve. It is easy to see that in the non-degenerated case where zero is met twice by the curve, the tangent approximates the curve with high order.*

$$R^2 \sin^2\left(\frac{\tau - \tau_1}{2}\right) - Rl(\cos\tau - \cos\tau_1) + l^2 = d^2. \qquad (2.11)$$

With (2.8) there holds:

$$x = \frac{R}{2}(\cos\tau + \cos\tau_1) = R\cos\left(\frac{\tau + \tau_1}{2}\right)\cos\left(\frac{\tau - \tau_1}{2}\right) \quad \text{and}$$

$$y = \frac{R}{2}(\sin\tau + \sin\tau_1) = R\sin\left(\frac{\tau + \tau_1}{2}\right)\cos\left(\frac{\tau - \tau_1}{2}\right),$$

hence

$$x^2 + y^2 = R^2 \cos^2\left(\frac{\tau - \tau_1}{2}\right). \qquad (2.12)$$

From this there follows

$$R^2(\cos\tau - \cos\tau_1)^2 = 4\sin^2\left(\frac{\tau + \tau_1}{2}\right)(R^2 - x^2 - y^2). \qquad (2.13)$$

Substituting

$$\sin^2\left(\frac{\tau + \tau_1}{2}\right) = \frac{y^2}{x^2 + y^2}$$

from (2.11) we get for x and y the algebraic equation of sixth degree

$$\boxed{\left[(R^2 + l^2 - d^2) - (x^2 + y^2)\right]^2 (x^2 + y^2) = 4l^2 y^2 (R^2 - x^2 - y^2).} \qquad (2.14)$$

This equation describes Watt's curve[24] for M.

Two examples of its graph are shown in figure 2.3 on page 29. The curve lies within the disc $x^2 + y^2 \leq R^2$. In the case $d = 1$ it touches the border.

---

[24] See [Gon47/1] for explanation of the choice of the name.

Equation (2.14) is of third degree in y^2, and so solvable in y^2.

In the non-degenerate case $(d + l > R)$ in a neighbourhood of the origin there hold for y the series expansions

$$\pm y = \left[\frac{R^2 + l^2 - d^2}{pq}\right] x + \left[8R^2 l^4 \frac{R^2 + l^2 - d^2}{p^5 q^5}\right] x^3 + \cdots \qquad (2.15)$$

with

$$p = \sqrt{(R + l)^2 - d^2} \qquad q = \sqrt{d^2 - (R - l)^2}. \qquad (2.16)$$

In the case $d + l = R$ this expansion is degenerated. Then we have expanding x

$$\pm x = \left[\frac{1}{R}\sqrt{\frac{R - l}{R}}\right] y^2 + \cdots . \qquad (2.17)$$

Comparing (2.15) and (2.17) we see that in the first case the curve is approximated by the tangent with order $O(x^2)$, in the second case with order $O(x)$. For $d^2 = l^2 - R^2$ the order will reach even fourth degree in the first case.

These ideas induced Chebyshev to deal with the following problem: Take two points P and Q on the curve with their vertical distances from the origin $-h$ and h. With the above-mentioned considerations the tangent well approximates the curve. Now Chebyshev intends to determine the parameters of the mechanism so that the maximal error of the approximation of the curve by the tangent **on the whole interval** $[-h, h]$ is minimized.

In January 1853 Chebyshev wrote down his first considerations about this problem in the work „Théorie des mécanismes, connus sous le nom de parallélogrammes" [Cheb54].

2.3 First Theoretical Approaches

The paper [Cheb54] can be counted as a scientific report about Chebyshev's trip abroad.

Stimulated by the new insights gained in England and France, Chebyshev tried to give mathematical foundations to the theory of mechanisms. As an introduction he mentioned that practical mechanics had not given suitable rules to find the mechanism with the smallest deviation from the ideal run. Until then parameters had been chosen so that the piston rod took a vertical position at the beginning, the middle and the end of a turn.[25] Already here

---

[25] Fig. 2.1 shows this position for the middle of one turn.

Chebyshev pointed out with the help of concrete examples that this method was not the best one (p. 112).[26]

In the second paragraph Chebyshev mentioned another method, the minimizing of the error function using Taylor's formula at the borders and the center. But this method could not help to solve the posed problem, since it was exact only in small neighbourhoods of those points, but not in a whole interval.

With these examples he could give good arguments for his new approach. He mentioned that it was only Poncelet who gave solutions for approximation problems of this kind[27] and explicitly cited his results on the above-mentioned approximation of square roots by linear expressions [Pon35]. Nevertheless Chebyshev then rejected his methods because the equations to determine the coefficients that Poncelet had found were untypically easy and only valid in special cases. Chebyshev wanted to solve more general problems.

At the end of the second paragraph he formulated for the first time the setting of the problem of best approximation (translated from the French word-by-word):

Problem: To determine the deviations which one has to add to get an approximated value for a function f, given by its expansion in powers of $x - a$, if one wants to minimize the maximum of these errors between $x = a - h$ and $x = a + h$, h being an arbitrarily small quantity.[28]

Of course the quantity h is disturbing. On the other hand it is not less surprising that Chebyshev assumes that it is analytic for the general case of a function to be approximated.

Already here we recognized a lack of this pioneering work: It can often be seen that Chebyshev here followed new paths—the text did not have the elegance of Chebyshev's later works, it was not smooth and left many questions open.

In paragraph 3 Chebyshev laid out general assumptions and considerations. The function f to be approximated is defined on the interval $[a-h, a+h]$, U being a polynomial of nth degree.

Immediately after these preparations Chebyshev stated for the number of deviation points:

---

[26] The page numbers are taken from volume 1 of the 1899 edition of Chebyshev's collected works [Chebgw1] by A. A. Markov and Sonin.

[27] [Cheb54], p. 114): „Relativement à la méthode d'approximation [...], nous n'avons que des recherches de M. Poncelet".

[28] [Cheb54], p. 114): „Déterminer les modifications qu'on doit apporter dans la valeur approchée de fx, donnée par son développement suivant les puissances de $x-a$, quand on cherche à rendre *minimum* la limite de ses erreurs entre $x = a - h$ et $x = a + h$, h étant une quantité peu considérable."

If we choose the coefficients of the polynomial "so that the difference $f(x) - U$ deviates from zero from $x = a - h$ to $x = a + h$ as little as possible, then the difference function $f(x) - U$, **as one knows**, satisfies the following property:

Between the largest and the smallest value $f(x) - U$ we find at least $n + 2$ points within the interval $[a - h, a + h]$ with the same values" (p. 114).[29]

Later in the paper he implicitly assumed that these points of equal value are also the deviation points. Unfortunately there is neither a proof nor a citation. Commentators can only guess, where this proposition came from, whether Chebyshev himself proved it or whether it had been a well-known fact which he became familiar with during his trip abroad.[30] Anyway it seems that Chebyshev did not want to prove it that time—the nature of this text was another one. Later Chebyshev caught up with the missing proofs of the properties of best approximating functions (in [Cheb59], written 1857). Since we do not know any sources that were written earlier than 1857, the assumption that Chebyshev was the first to find a formal proof for this proposition is the most probable.

In the following we summarize Chebyshev's theoretical results from [Cheb54].

2.3.1 Characteristic Equations

As mentioned above, Chebyshev restricted the problem to real analytic functions approximated by polynomials of nth degree on an interval $[a - h, a + h]$. Let f be the given function, P the polynomial to be computed and

$$L := \max_{a - h \le x \le a + h} |f(x) - P(x)|$$

the maximal error of approximation.

With f real analytic on $[a - h, a + h]$ and the "well-known" fact that there are at least $n + 2$ deviation points we can set the following $2n + 4$ characteristic equations which hold in the deviation points of $f - P$:

$$(x - a + h)(x - a - h)[f'(x) - P'(x)] = 0 \qquad (2.18)$$
$$L^2 - [f(x) - P(x)]^2 = 0.$$

Now Chebyshev attempted to solve this problem for some special cases.

---

[29] [Cheb54], S. 114: „Si l'on choisit ces coefficients de manière à ce que la différence $fx - U$, depuis $x = a - h$, jusqu'à $x = a + h$, reste dans les limites les plus rapprochées de 0, la différence $fx - U$ jouira, **comme on le sait**, de cette propriété: Parmi les valeurs les plus grandes et les plus petites de la différence $fx - U$ entre les limites $x = a - h$, $x = a + h$, on trouve au moins $n + 2$ fois la même valeur numérique". [Emphasis by the author].

[30] Comp. [Gon45], p. 128.

2.3.2 Approaches for Real-Analytic Functions

With the definition

$$k_i := \frac{f^{(i)}(a)}{i!} \qquad (2.19)$$

and the transformation $z := \frac{x-a}{h}$ the problem is reformulated for the first time ([Cheb54], p. 116). With modern words the problem was reduced to that of determining the polynomial $P(h, z)$ defined on the interval $[-1, 1]$ which deviates as little as possible from f with

$$f(h, z) := \sum_{i=0}^{\infty} k_i (hz)^i. \qquad (2.20)$$

Here P is of degree n both in h and in z.

Chebyshev now simplified the problem with the following restrictions:

- Consider the number m of Taylor coefficients of f, vanishing directly after the nth Taylor coefficient (that is, $k_{n+1} = \ldots = k_{n+m} = 0$).
- Solve the problem neglecting an error of order $O(h^{n+m+1})$.

According to these facts he could state:

1. The best approximating polynomial of order $O(h^{n+1})$ is the nth partial sum of the Taylor series of f (p. 116).[31]
2. The problem now is reduced to the determination of the polynomial of $(n + m + 1)$th degree with given first coefficient which deviates on the interval $[-1, 1]$ as little as possible from zero (neglecting an error of order $O(h^{m+n+1}))$,[32] that means that it is to determine a polynomial T_{n+m+1}, with degree $n + m + 1$ and

---

[31] Chebyshev did not prove this. It can be done easily: Let $S := \sum\limits_{i=0}^{n} k_i \iota^i$. Then we have for S and the best approximating polynomial $P(h, .)$

$$\max_{|z| \leq 1} |S(hz) - f(hz)| \leq C_1 h^{n+1} \quad \text{and}$$

$$\max_{|z| \leq 1} |P(h, z) - f(hz)| \leq C_2 h^{n+1}$$

for real constants C_1 and C_2 and arbitrarily small $h > 0$. So we also have

$$\max_{|z| \leq 1} |P(h, z) - S(hz)| = O(h^{n+1}),$$

and because P and S have the same degree in h, finally

$$S \equiv P$$

[32] With the above-mentioned considerations we can split $P(h, z)$ into

$$\max_{|z|\leq 1}|\sigma T_{n+m+1}(z)| \quad \text{minimal}, \quad \sigma \in \mathbb{R}. \tag{2.21}$$

2.3.3 The Polynomial of $(n + 1)$thDegree and Given First Coefficient Least Deviating from Zero

Again Chebyshev restricted the problem. He remarked (p. 121) that only the case $m = 0$ has a practical meaning for the theory of mechanisms and for it he gave a general solution of the (restricted) problem.

The characteristic equations now have the form

$$(z^2 - 1)\frac{d(\sigma z^{n+m+1} - Q_0(z))}{dz} = 0 \quad \text{and} \tag{2.22}$$
$$L^2 - [(\sigma z^{n+m+1} - Q_0(z))] = 0.$$

We abbreviate $y(z) := \sigma z^{n+m+1} - Q_0$. Using the fact that these equations have $n + 2$ common roots, Chebyshev showed that for $z \in [-1, 1]$,

$$\frac{[y(z)]^2 - L^2}{[y'(z)]^2} = \frac{P(z)(z^2 - 1)}{[Q(z)]^2}, \quad \text{with } \deg P = 2m \quad \text{and } \deg Q = m. \tag{2.23}$$

Finally we get the differential equation

$$\frac{dy}{\sqrt{[y(z)]^2 - L^2}} = \frac{Q(z)dz}{\sqrt{P(z)(z^2 - 1)}}, \quad |z| > 1, \tag{2.24}$$

which can be solved with methods Chebyshev derived in his dissertation pro venia legendi "On Integration with the help of logarithms" [Cheb47] based on calculations developed by Abel.[33]

$$P(h, z) = S(hz) + h^{n+m+1}Q(z, h),$$

where $Q(., h)$ is a polynomial of degree n in z. So we split again:

$$Q(z, h) = Q_0(z) + hQ_1(x) + \ldots + h^l Q_l(z) + O(h^{l+1})$$

with $\deg Q_i = n \quad \forall i = 0, \ldots, l$. Now we compute

$$\max_{|z|\leq 1}|P(h, z) - f(hz)| = \max_{|z|\leq 1}|\sum_{i=n+m+1}^{\infty} k_i h^i x^i - h^{n+m+1}Q(z, h)|$$
$$= h^{n+m+1}\max_{|z|\leq 1}|k_{n+m+1}x^{n+m+1} - Q_0(z) + O(h)|.$$

Since $P(., h)$ is the best approximating polynomial on $[-1, 1]$ for f, all these expressions will become as small as possible.

[33] Chebyshev cited a letter from Abel to Legendre [Chebgw2, Vol. 5, p. 90].

2.3.3.1 Determination of T_n. The Case $m = 0$

For $m = 0$ the polynomials P and Q become constants.

The differential equation to be solved is

$$\frac{dy}{\sqrt{[y(z)]^2 - L^2}} = \lambda \frac{dz}{\sqrt{(z^2 - 1)}} \quad \text{with a constant } \lambda. \qquad (2.25)$$

After integration we get

$$\lambda \log \frac{z + \sqrt{z^2 - 1}}{z - \sqrt{z^2 - 1}} + C = \log \frac{y + \sqrt{y^2 - L^2}}{y - \sqrt{y^2 - L^2}}. \qquad (2.26)$$

It is easy to derive from (2.22) that there must be extremal points at the borders of the interval ($z = \pm 1$). Therefore we have $C = 0$, and y can be calculated.

Equation (2.26) implies that

$$\frac{y + \sqrt{y^2 - L^2}}{y - \sqrt{y^2 - L^2}} = \frac{(z + \sqrt{z^2 - 1})^\lambda}{(z - \sqrt{z^2 - 1})^\lambda}. \qquad (2.27)$$

With $z^+ := (z + \sqrt{z^2 - 1})^\lambda$ and respectively $z^- := (z - \sqrt{z^2 - 1})^\lambda$ we have $z^+ z^- \equiv 1$, and equation (2.27) leads to

$$yz^- + \sqrt{y^2 - L^2}z^- = yz^+ - \sqrt{y^2 - L^2}z^+, \quad \text{hence}$$

$$1 - \frac{L^2}{y^2} = \left(\frac{z^+ - z^-}{z^+ + z^-}\right)^2, \quad \text{and finally}$$

$$y = \pm\frac{L}{2}[(z + \sqrt{z^2 - 1})^\lambda + (z - \sqrt{z^2 - 1})^\lambda]. \qquad (2.28)$$

For $m = 0$ this is the polynomial of degree $(n+1)$ with the $(n+1)$th coefficient k_{n+1} which, fixed on the interval $[-1, 1]$, deviates the least possible from zero.[34]

Afterwards Chebyshev calculated the value of the exponent λ and the maximal error L. Since we assumed $m = 0$, we have $\deg y = n+1$, so $\lambda = n+1$. With k_{n+1} given, there holds[35]

---

[34] This is the function which at least solves equation (2.25) for $|z| > 1$. It has to be proved whether this function is defined for $z \in [-1, 1]$. Indeed all expressions under the roots vanish using the binomial formula, since

$$(z + \sqrt{z^2 - 1})^\lambda + (z - \sqrt{z^2 - 1})^\lambda = 2 \sum_{k=0, k \text{ even}}^{\lambda} \binom{\lambda}{k} z^{\lambda-k}(z^2 - 1)^{\frac{k}{2}}.$$

[35] Since we now know that the use of the root expressions is valid, we calculate applying the binomial formula:

$$L = \pm \frac{k_{n+1}}{2^n}.$$

So Chebyshev solved the approximation problem for $f(x) = \sigma x^{n+1}$. In the next paragraphs the solutions for any m are determined recursively and the problem is solved for $f(x) = x^{n+m+1}$.

Step by step similar approaches are used there for the polynomials Q_i from equation (32), and the order of the error can be scaled down.

2.3.4 Remarks on „Théorie des mécanismes..."

Indeed this paper is not as elegant as Chebyshev's usual publications mainly are. It is more a provisionary arrangement which introduced two of Chebyshev's main subjects of interest: mechanism and approximation theory. The latter did not have its own name at that time. Later it would be called "theory of functions deviating the least possible from zero." A. A. Markov would name a whole lecture with this monstrous expression ([MarA06], see also section 3.6). Possibly the text had to answer the purpose of a scientific report about the trip abroad. This might have caused a pressure of time and might be an explanation for the present style and the missing end of the paper— Chebyshev finished the text with a reference to supposed following paragraphs, where he intended to show applications of his calculations to mechanism theory.[36] These paragraphs, however, have never existed. Chebyshev decided to go another way, in fact he would write a large number of papers where he investigated certain properties of mechanisms or developed and even constructed them. After four years he would return to the theory with his above-mentioned voluminous monograph „Sur les questions de minima [...]" [Cheb59].

One of the most remarkable passages of „Théorie des mécanismes [...]" is the place where Chebyshev cited the statement about the number of deviation points, but he neither proved it nor gave a source for another interesting point to look at. Although Chebyshev aimed to investigate all formulae with respect to all implications he did not mention the alternation property, although it easily follows from the characteristic equations (2.22). Substituting $y(z) := \sigma z^{n+m+1} - Q_0(z)$ for the case $m = 0$ they have the form

$$(z + \sqrt{z^2 - 1})^\lambda = \sum_{k=0}^{\lambda} \binom{\lambda}{k} z^k \left(\sum_{i=0}^{\lambda-k} \binom{\lambda-k}{i} z^{2i} (-1)^{\lambda-k-i} \right)^{1/2}.$$

We state:

- Neglecting the expressions of type $(-1)^{\lambda-k-i}$ we see that z^λ has the coefficient $\sum \binom{\lambda}{k} = 2^\lambda$
- To $(z + \sqrt{z^2 - 1})^\lambda$ there exists a corresponding expression $(z - \sqrt{z^2 - 1})^\lambda$, multiplied with $(-1)^{\lambda-k-i+1}$, so then the sum will be zero. So the highest coefficient of y is $\pm 2^{\lambda-1} L$.

[36] [Cheb54], S. 143: „Dans les §§ suivants nous montrerons l'usage des formules."

$$(z^2 - 1)\frac{dy}{dz} = 0 \quad \text{and}$$
$$L^2 - [y(z)]^2 = 0.$$

Since y is a polynomial of degree $n + 1$, it can't have more than n extrema. Because there are $n + 2$ deviation points, all inner deviation points are extrema. If they did not alternate in signs, between two of them we had another extremum!

Although the alternation property directly follows, Chebyshev did not mention it. But it did not help to calculate the best approximating polynomial, so it had only restricted meaning for the aim he followed with his pioneering text.

The same statement holds for questions regarding existence and uniqueness of the problem. These questions were not of an a priori importance for him since he found a unique solution for a special case.

2.4 First Theoretical Compositions

Four years later Chebyshev presented the monograph ,,Sur les questions des minima qui se rattachent à la représentation approximative des fonctions" [Cheb59][37] before the Imperial Academy of Sciences. It reached much farther than the ,,Théorie des mécanismes ..." discussed in the last paragraph. Here we find questions of a generally theoretical interest, e. g., a necessary criterion for best approximating functions, a proof of the proposition about the number of deviation points and also concrete solutions of three special problems of approximation.

2.4.1 Questions about Minima

The style of this monograph immediately attracts the attention of the reader. It is written like a textbook and can be read without any knowledge of the predecessor ,,Théorie des mécanismes ...". All results and all problems are newly arranged.

Again Chebyshev emphasized here that he dealt with a new form of approximating functions, for instance basically differing from Taylor polynomials.

But the new and wider approach allowed arbitrary intervals and approximations independent from the quantity h, which stood for the interval length.

The problem he posed is far more general than that of the 'mechanism theory':

---

[37] The page numbers are again taken from [Chebgw1, Vol. 1].

Let $F(x)$ be an arbitrary but fixed function with certain parameters $p_1, p_2, \ldots, p_n$. Now determine the values of $p_1, p_2, \ldots, p_n$ so that the deviation of F on an interval $[-h, h]$ becomes as small as possible.[38]

So Chebyshev tried to solve the following minimax-problem:

$$\max_{p_1, \ldots, p_n \in \mathbb{R}, x \in [-h, h]} |F(x, p_1, \ldots, p_n)| \to \min. \qquad (2.29)$$

He extended the problem from the original problem, the approximation by polynomials to a wider class which included, e. g., trigonometric and rational functions. The general formulation shows that he did not want to restrict the problem unnecessarily, although he soon would limit his considerations to polynomial, weighted and rational approximation.

2.4.2 A General Necessary Criterion

For the following considerations Chebyshev assumed F to be differentiable according to all variables and parameters and its partial derivatives $\frac{\partial F}{\partial p_i}$ to be bounded [Cheb59, p. 276, §6].

Then he repeated setting the characteristic equations which now are

$$(x^2 - h^2)F'(x) = 0 \quad \text{and} \qquad (2.30)$$
$$F^2(x) - L^2 = 0.$$

With these settings Chebyshev was able to propose an important basic necessary criterion for the solution of the above-described approximation problem with n parameters:

Theorem 2.1 (Chebyshev's Theorem) *$F : [-h, h] \times \mathbb{R}^n \to \mathbb{R}$ is not the best approximation of the zero function, if the linear system of equations*

$$\lambda_1 \frac{\partial F}{\partial p_1}(x_1) + \lambda_2 \frac{\partial F}{\partial p_1}(x_2) + \ldots + \lambda_\mu \frac{\partial F}{\partial p_1}(x_\mu) = 0$$

$$\lambda_1 \frac{\partial F}{\partial p_2}(x_1) + \lambda_2 \frac{\partial F}{\partial p_2}(x_2) + \ldots + \lambda_\mu \frac{\partial F}{\partial p_2}(x_\mu) = 0 \qquad (2.31)$$

$$\vdots \qquad\qquad \vdots \quad \vdots$$

$$\lambda_1 \frac{\partial F}{\partial p_n}(x_1) + \lambda_2 \frac{\partial F}{\partial p_n}(x_2) + \ldots + \lambda_\mu \frac{\partial F}{\partial p_n}(x_\mu) = 0$$

is only solved by the trivial solution $\lambda_1 = \ldots = \lambda_\mu = 0$. The points x_i, $i = 1, \ldots, \mu$ here are the (only) deviation points of $F(., p_1, \ldots, p_n)$.

---

[38] [Cheb59, S. 274:] „*Etant donnée une fonction quelconque $F(x)$ avec n paramètres arbitraires $p_1, p_2, \ldots, p_n$, il s'agit par un choix convenable des valeurs $p_1, p_2, \ldots, p_n$ de rendre minimum la limitie de ses écarts de zéro entre $x = -h$ et $x = +h$.*"

At first we state that Chebyshev did not propose a fixed number of deviation points. But he (theoretically) restricted the problem to the case of having a finite number of deviation points, as we see both in the formulation and later in the proof. In fact Chebyshev's concrete problems would satisfy this condition, so he might have regarded an infinite number of deviation points as degenerated.

The theorem later would help to compute the number of deviation points, μ. We note that the linear system of equations (2.31) has only the trivial solution, if the number of columns is less than or equal to the number of rows, or if there holds

$$\mu \leq n.$$

This implies the following statement, which Chebyshev proved of course without arguments from modern linear algebra.[39] He showed that there exists a non-trivial solution $(N_1, \ldots, N_n) \in \mathbb{R}^n$ of the following linear system of equations:

$$N_1 \frac{\partial F}{\partial p_1}(x_1) + N_2 \frac{\partial F}{\partial p_2}(x_1) + \ldots + N_n \frac{\partial F}{\partial p_n}(x_1) = F(x_1)$$

$$N_1 \frac{\partial F}{\partial p_1}(x_2) + N_2 \frac{\partial F}{\partial p_2}(x_2) + \ldots + N_n \frac{\partial F}{\partial p_n}(x_2) = F(x_2) \qquad (2.32)$$

$$\vdots \qquad\qquad \vdots \qquad\qquad \vdots$$

$$N_1 \frac{\partial F}{\partial p_1}(x_\mu) + N_2 \frac{\partial F}{\partial p_2}(x_\mu) + \ldots + N_n \frac{\partial F}{\partial p_n}(x_\mu) = F(x_\mu).$$

If there exists a non-trivial solution for (2.32), then the function $F(., p_1, \ldots, p_n)$ approximating the zero function can be improved. Such an improvement could be

$$F_0(x) = F(x, p_1, \ldots, p_n) - \left(\sum_{j=1}^n \frac{\partial F}{\partial p_j}(x) N_j \right) \omega + o(\omega)_{\omega \to 0}$$

with a small ω. To illustrate Chebyshev's arguments we want to translate them into more modern mathematics:

We have for $p = (p_1, \ldots, p_n)$, $N = (N_1, \ldots, N_n)$, $\theta \in (0,1)$ and a fixed x:

$$F_0(x,p) = F(x, p - \omega N)$$
$$= F(x,p) - <\mathrm{grad} F(x, p - \theta\omega N), \omega N> + o(\theta\omega\|N\|).$$

For x_i the deviation points there holds:

$$F_0(x_i, p) = F(x_i, p) - \omega \sum_{j=1}^n [N_j \frac{\partial F}{\partial p_j}(x_i, p) + C_1] + o(\omega\|N\|)$$

---

[39] Chebyshev proved the above-mentioned (general) evidence of the rank of a matrix for this special case. (p. 278 f.).

with a constant C_1. Using the system of equations (2.32) we get

$$F_0(x_i, p) = (1 - \omega)F(x_i, p) - \omega C_2,$$

C_2 being a constant. So, with a suitable choice of ω we can improve F in the deviation points and so in small neighbourhoods of them, due to the continuity of F. Since the partial derivatives $\frac{\partial F}{\partial p_i}$ are bounded, the equation determining F_0 shows that ω can be suitably chosen that beyond these neighbourhoods the function can be improved, too.

This construction is only possible, however, if there exists only a finite number of deviation points of $F(., p_1, \ldots, p_n)$. Chebyshev did not discuss this fact, as mentioned above. How to avoid this problem was shown much later in [Nat49, p. 27 ff.].

This theorem set the foundations for a list of concrete solutions and corollaries.

2.4.3 The Number of Deviation Points. Three Cases

Chebyshev introduced the following paragraphs (§§9-12) remarking that the number of deviation points was generally unknown. He stated that

- in the case $\mu > n$ the characteristic equations (2.30) suffice to determine the unknown variables.
- in the case $\mu \leq n$ we get, after solving (2.31), $n - \mu + 1$ equations of the parameters $p_1, \ldots, p_n$ and $x_1, \ldots, x_\mu$. Together with the 2μ equations (2.30) we have again a sufficient number of equations.
- if there is no information about μ, we have to determine μ step by step using the computed value for L.

The further considerations were restricted to three cases:

1. Approximation by polynomials:

$$F(x, p) = f(x) - \sum_{i=1}^{n} p_i x^{i-1}. \tag{2.33}$$

2. Approximation by polynomials using a polynomial weight:

$$F(x, p) = f(x) - \frac{1}{Q(x)} \sum_{i=1}^{n} p_i x^{i-1}, \quad Q \text{ Polynom.} \tag{2.34}$$

3. Rational approximation:

$$F(x, p) = f(x) - \left(\sum_{i=1}^{n-l} p_i x^{i-1} \right) \bigg/ \left(\sum_{i=1}^{l} p_{n-l+i} x^{i-1} \right). \tag{2.35}$$

Here each f is differentiable on $[-h, h]$.

We want to show his arguments in the case of the polynomial approximation. The second case is analogous; in the third case we have to have in mind the specific discontinuous behaviour of rational approximation where, for example, the number of equations of (2.32) depends on the degree of the best approximation after cancellation.

It is interesting to notice that except in this paragraph there is no other statement on non-linear approximation, either by Chebyshev or by other authors mentioned in the present book. The only exception is Kirchberger's thesis [Kir02].

2.4.3.1 Approximation by Polynomials

In this case the matrix of the system of equations (2.31) is Vandermond's matrix

$$V_{n,\mu}(x_1, \ldots, x_\mu) = \begin{pmatrix} x_1^{n-1} & x_2^{n-1} & \cdots & x_\mu^{n-1} \\ x_1^{n-2} & x_2^{n-2} & \cdots & x_\mu^{n-2} \\ \cdots\cdots\cdots\cdots\cdots \\ x_1 & x_2 & \cdots & x_\mu \\ 1 & 1 & \cdots & 1 \end{pmatrix}. \tag{2.36}$$

This matrix has rank μ for $\mu \leq n$ and so the linear system of equations (2.31)

$$V_{n,\mu}(x_1, \ldots, x_\mu)(\lambda_1, \ldots, \lambda_\mu)^T = 0$$

has only the trivial solution, which would be impossible for the best approximation. So for the best approximation we have:

$$\mu \geq n + 1. \tag{2.37}$$

Thus, we proved the statement Chebyshev cited as 'well-known' in the monograph ,,Théorie des mécanismes''.[40]

Chebyshev formulated this fact in connection with the characteristic equations (2.30) as a theorem (Théorème 2, p. 284).

In the case of weighted polynomial approximation the theorem was proved analogously. The weight is not more than a factor before the matrix, having no influence on the number of deviation points (Théorème 3, p. 287).

Now Chebyshev returned to a well-known problem and started to determine concrete best approximating functions, to be precise a polynomial, a weighted polynomial and a rational function of degree n with given first coefficient 'least deviating from zero'.

In the following we will explain his solution for the polynomial case.

---

[40] Note that we reduced the number of parameters from n to $(n-1)$. Therefore the number of deviation points had to be reduced, too.

2.4.4 Determination of T_n

Now the aim was to determine parameters $p_1, \ldots, p_n$ so that the polynomial

$$F(x) := x^n + p_1 x^{n-1} + \cdots + p_n \quad (x \in [-h, h]) \tag{2.38}$$

is minimal according to the norm $\|.\|_\infty$. Chebyshev mentioned that this was obviously problem (2.33) for $f(x) = x^n$.

We recently determined that the error function (2.38) has at least $n + 1$ deviation points $x_0, \ldots, x_n$, all different from each other and solving the equations (2.30).

So it is clear that the expression

$$(x^2 - h^2)\left(F^2(x) - L^2\right)$$

can be divided by all the $2n + 2$ factors

$$(x - x_0)^2, (x - x_1)^2, \ldots, (x - x_n)^2.$$

Hence we have

$$(x^2 - h^2)\left(F^2(x) - L^2\right) = C(x - x_0)^2 (x - x_1)^2, \ldots, (x - x_n)^2 \tag{2.39}$$

with a positive constant C. Now it is clear that $-h$ and h must be deviation points, too. Without a restriction we set

$$x_0 := -h, \quad x_1 := h$$

and define

$$\Phi(x) := \sqrt{C} \prod_{i=2}^{n}(x - x_i). \tag{2.40}$$

So equation (2.39) becomes

$$\left(F^2(x) - L^2\right) = (x^2 - h^2)\Phi^2(x). \tag{2.41}$$

This equation can be written as

$$\left(F(x) - \Phi(x)\sqrt{x^2 - h^2}\right)\left(F(x) + \Phi(x)\sqrt{x^2 - h^2}\right) = L^2,$$

and we get

$$\frac{F(x)}{\Phi(x)} = \sqrt{x^2 - h^2} + \frac{L^2}{\Phi(x)\left(F(x) + \Phi(x)\sqrt{x^2 - h^2}\right)}. \tag{2.42}$$

Since F has degree n, Φ is of degree $n - 1$ and they have no common roots, we see from this representation that the quotient describes the continuous fraction of nth degree of the expansion of $\sqrt{x^2 - h^2}$.

If we name it by

$$\frac{P_n}{Q_n},$$

then we have with (2.42):

$$F = C_0 P_n \qquad \Phi = C_0 Q_n \tag{2.43}$$

with a constant C_0.

2.4.4.1 Expansion of continous fractions

From the equation

$$h^2 = (x - \sqrt{x^2 - h^2})(x + \sqrt{x^2 - h^2}) \tag{2.44}$$

we get

$$\sqrt{x^2 - h^2} - x = -\frac{h^2}{x + \sqrt{x^2 - h^2}}$$

$$= -\frac{h^2}{2x + \sqrt{x^2 - h^2} - x},$$

and this can be developed furthermore into

$$\sqrt{x^2 - h^2} - x = -\frac{h^2}{2x - \dfrac{h^2}{2x + \sqrt{x^2 - h^2} - x}}.$$

This leads to the continuous fraction

$$\sqrt{x^2 - h^2} = x - \cfrac{h^2}{2x - \cfrac{h^2}{2x - \cfrac{h^2}{2x - \cfrac{h^2}{2x - \cfrac{h^2}{2x - \dots}}}}}, \qquad \text{and} \tag{2.45}$$

using any finite representation we get the denominator $x + \sqrt{x^2 - h^2}$. If we name its corresponding fractions by $\frac{P_1}{Q_1}, \frac{P_2}{Q_2}, \dots, \frac{P_n}{Q_n}$, we get, using a fundamental theorem on continuous fractions[41]:

$$\sqrt{x^2 - h^2} = \frac{P_m\left(\sqrt{x^2 - h^2} + x\right) - h^2 P_{m-1}}{Q_m\left(\sqrt{x^2 - h^2} + x\right) - h^2 Q_{m-1}}. \tag{2.49}$$

---

[41] comp. e. g. [MarA48, p. 293 f.]: Let the continuous fraction

After some transformations using equation (2.44) we get

$$x - \sqrt{x^2 - h^2} = \frac{P_m - Q_m\sqrt{x^2 - h^2}}{P_{m-1} - Q_{m-1}\sqrt{x^2 - h^2}}. \tag{2.50}$$

Successively substituting m by $1, \ldots, n$ we get

$$\frac{P_n - Q_n\sqrt{x^2 - h^2}}{P_{n-1} - Q_{n-1}\sqrt{x^2 - h^2}} = x - \sqrt{x^2 - h^2}$$

$$\frac{P_{n-1} - Q_{n-1}\sqrt{x^2 - h^2}}{P_{n-2} - Q_{n-2}\sqrt{x^2 - h^2}} = x - \sqrt{x^2 - h^2}$$

$$\vdots = \vdots$$

$$\frac{P_2 - Q_2\sqrt{x^2 - h^2}}{P_1 - Q_1\sqrt{x^2 - h^2}} = x - \sqrt{x^2 - h^2}.$$

Multiplying all equations this leads to

$$\frac{P_n - Q_n\sqrt{x^2 - h^2}}{P_1 - Q_1\sqrt{x^2 - h^2}} = \left(x - \sqrt{x^2 - h^2}\right)^{n-1}. \tag{2.51}$$

For $P_1 = x$ and $Q_1 = 1$ the result is

$$P_n - Q_n\sqrt{x^2 - h^2} = \left(x - \sqrt{x^2 - h^2}\right)^n. \tag{2.52}$$

It is easy to check that the sign did not play a decisive role. The same calculations hold for $-\sqrt{x^2 - h^2}$.) Therefore we also have

$$P_n + Q_n\sqrt{x^2 - h^2} = \left(x + \sqrt{x^2 - h^2}\right)^n. \tag{2.53}$$

Connecting the equations (2.52) and (2.53) we have

$$\cfrac{a_1}{b_1 + \cfrac{a_2}{b_2 + \cfrac{a_3}{b_3 + \ldots}}}$$

be given. Then we have for the first three of the analogous expressions P_i, Q_i immediately

$$P_1 = a_1 \quad P_2 = a_1 b_2 \qquad P_3 = a_1 b_2 b_3 + a_1 a_3 \tag{2.46}$$

$$Q_1 = b_1 \quad Q_2 = b_1 b_2 + a_2 \quad Q_3 = (b_1 b_2 + a_2)b_3 + b_1 a_3, \tag{2.47}$$

and there holds the recursion formula

$$P_k = b_k P_{k-1} + a_k P_{k-2}; \quad Q_k = b_k Q_{k-1} + a_k Q_{k-2} \tag{2.48}$$

$$P_n = \frac{\left(x + \sqrt{x^2 - h^2}\right)^n + \left(x - \sqrt{x^2 - h^2}\right)^n}{2} \quad \text{and} \quad (2.54)$$

$$Q_n = \frac{\left(x + \sqrt{x^2 - h^2}\right)^n - \left(x - \sqrt{x^2 - h^2}\right)^n}{2\sqrt{x^2 - h^2}}. \quad (2.55)$$

The validity of these expressions has already been discussed in section 2.3.3.1 on page 35.

With (2.43) we have

$$F = C_0 \frac{\left(x + \sqrt{x^2 - h^2}\right)^n + \left(x - \sqrt{x^2 - h^2}\right)^n}{2} \quad \text{(comp. (2.38))}.$$

Since the leading coefficient of F was set equal to 1 and P_n has the leading coefficient 2^{n-1}[42], we get

$$C_0 = \frac{1}{2^{n-1}}$$

and

$$F = \frac{1}{2^n}\left[\left(x + \sqrt{x^2 - h^2}\right)^n + \left(x - \sqrt{x^2 - h^2}\right)^n\right] = T_n. \quad (2.56)$$

The maximal error is then

$$L := \|F\|_\infty = F(h) = \frac{h^n}{2^n}. \quad (2.57)$$

2.4.5 Solvability of Algebraic Equations

Based on these results Chebyshev derived a list of corollaries describing the places zeros of algebraic equations may take.

At first he stated a tricky, but simple consequence which follows from the results above:

Corollary 2.2 [Cheb59, Theorem 5] *Let p be a polynomial of degree n on $[-h, h]$ with the following shape:*

$$p(x) := x^n + \sum_{i=1}^{n-1} a_i x^i.$$

Then

$$\|p\|_\infty \geq 2\left(\frac{h}{2}\right)^n. \quad (2.58)$$

With the help of this he easily showed a necessary condition for the monotonicity of a polynomial of such a shape:

---

[42] comp. the calculation in paragraph 2.3.3.1 on page 35.

Corollary 2.3 [Cheb59, Theorem 6] *Let p be a polynomial of degree n on $[-h, h]$ with*

$$p(x) := x^n + \sum_{i=1}^{n-1} a_i x^i.$$

If additionally

$$|p(-h) - p(h)| < 4 \left(\frac{h}{2} \right)^n,$$

then p is on $[-h, h]$ neither strictly monotonic decreasing nor strictly monotonic increasing.

Otherwise the polynomial shifted by a respectively chosen constant would have a norm less than $2 \left(\frac{h}{2} \right)^n$.

It is possible to get information about the points where p changes its sign:

Corollary 2.4 [Cheb59, Theorem 8] *Let p be a polynomial of degree n on $[-h, h]$ with*

$$p(x) := x^n + \sum_{i=1}^{n-1} a_i x^i.$$

Then there holds for all $t \in [-h, h]$:

Between the points t and $t - 4 \operatorname{sign} \frac{p(t)}{p'(t)} \sqrt[2n]{\frac{p^2(t)}{16}}$ there is a point, where p and p' have opposite signs.

So we can conclude information about the places of the zeros of monotone polynomials of this shape, especially for polynomials having only odd non-vanishing powers. Chebyshev proved the following theorem:

Theorem 2.5 [Cheb59, Theorem 11] *The equation*

$$x^{2l+1} + \sum_{i=0}^{l-1} a_i x^{2i+1} = 0 \tag{2.59}$$

has at least one root within the interval

$$\left[-2 \sqrt[2l+1]{\frac{a_0}{2}}, 2 \sqrt[2l+1]{\frac{a_0}{2}} \right].$$

Statements of this were of course very important for the approximate solution of non-solvable algebraic equations.

Later he would improve this estimate with respect to monotone polynomials of least deviation from zero. We will return here in section 2.5.4.

2.4.6 Application to Interpolation Problems

In paragraph VIII Chebyshev described an interesting application of his results. He was able to improve the error of Lagrange's interpolation formula by means of T_n.

Let $f \in C^n[-h, h]$ be a function.[43] to be interpolated in knots $x_1, \ldots, x_n \in [-h, h]$ by a polynomial of degree $n - 1$,

$$p(x) := \sum_{i=0}^{n} a_i x^i,$$

As well known, the error of interpolation is

$$f(x) - p(x) = \frac{f^{(n)}(\xi)}{n!} \prod_{i=1}^{n} (x - x_i), \quad \xi \in (a, b). \tag{2.60}$$

Consequently Chebyshev noticed that it mainly depends on the choice of the knots. Using this he now tried to improve the error of interpolation independently on the shape of f.

He knows the polynomial of least deviation from zero (with fixed first coefficient). The approach

$$\prod_{i=1}^{n} (x - x_i) = T_n(x) \tag{2.61}$$

now leads to the error

$$\|f - p\|_\infty = \left| \frac{f^{(n)}(\xi)}{n!} \right| \|T_n\|_\infty = \left| \frac{f^{(n)}(\xi)}{n!} \right| \frac{h^n}{2^{n-1}}, \tag{2.62}$$

which is minimal.

So the sought knots are the zeros of T_n,

$$x_i = h \cos \frac{(2i - 1)\pi}{2n}, \quad i = 1, \ldots, n. \tag{2.63}$$

With the help of Chebyshev's result it is clear that the sequence of Lagrange interpolation polynomials converges for any real-analytic function f defined on an interval of length $h < 2$, if only the knots are chosen in the above-described way.

Although obvious, this result was not mentioned by Chebyshev, and so in 1904 it was Carl Runge, who first published this result (in [Run04, S. 136 ff.].[44])

---

[43] Chebyshev demanded that $f, f', \ldots, f^{(n)}$ should be continuous and bounded on an interval covering the knots $x_1, \ldots, x_n$ ([Chebgw1, S. 309]: „...tant que la fonction $f(x)$ et ses dérivées $f'(x), f''(x), \ldots, f^{(n-1)}(x), f^{(n)}(x)$ ne cessent d'être finies et continues dans les limites où sont comprises $x, x_1, \ldots, x_n$ [...]").

[44] Cited after [Ric85, S. 163].

Runge's biographer Richenhagen [Ric85] points out that Runge obviously did not know Chebyshev's results because he had even to calculate the Chebyshev polynomials T_n. So it is clear that Runge was not able to mention this result.[45]

2.4.7 Evaluation of 'Questions about Minima'

„Sur les questions des minimas [...]" [Cheb59] is a theoretical work, where Chebyshev laid out the foundations of modern approximation theory. He stated a very general problem (2.29) enormously abstracting from concrete questions. Especially Théorème 1 giving a necessary criterion for the solution of (2.29) is a real classic.

Consider the standard case, the approximation of continuous[46] functions by polynomials of fixed degree n.

With $x_1, \ldots, x_\mu$ as deviation points of the minimal solution, we get from Chebyshev's theorem a non-trivial representation of zero, to be precise

$$0 = \sum_{i=1}^{\mu} \lambda_i \begin{pmatrix} 1 \\ x_i \\ x_i^2 \\ \vdots \\ x_i^{n-1} \end{pmatrix}. \tag{2.64}$$

Multiplying this equation with the values $F(x_i)$ of the deviation points, we get

$$0 = \sum_{i=1}^{\mu} \hat{\lambda}_i F(x_i) \begin{pmatrix} 1 \\ x_i \\ x_i^2 \\ \vdots \\ x_i^{n-1} \end{pmatrix}, \tag{2.65}$$

where $\hat{\lambda}_i$ vanish only, if λ_i vanish. If it is possible to prove that all the $\hat{\lambda}_i$ are non-negative numbers, then a citerion of the type 'zero in the convex hull' (see, e. g. [Che66]) will be found, which is important in optimization theory and usually cited as a theorem of Kuhn and Tucker. Kirchberger was the first to prove it [Kir02].

The generality of Chebyshev's theorem is remarkable. Meinardus and Schwedt [MeSc64] developed the theory of approximating functions which

---

[45] Since Richenhagen does not mention it, either, it also seems to be unknown to him. His remarks only regard the definition and properties of the T_n, not Runge's theorem on the convergence of the Lagrange interpolation formula.

[46] Although Chebyshev demanded F to be differentiable, it is only necessary to have partial differentiability according to the parameters $p_1, \ldots, p_n$.

are differentiable by parameters $p_1, \ldots p_n$ based on Chebyshev's theorem and including rational approximation and approximation by exponential sums. Central for this theory is the concept of a tangential manifold. They are generated by the gradients

$$\left(\frac{\partial F}{\partial p_1}(x_i), \ldots, \frac{\partial F}{\partial p_n}(x_i) \right),$$

calculated in the points $x_1, \ldots, x_\mu$, that is, the columns of (2.31).

The alternation theorem, often named after Chebyshev, also originates from this theorem. In addition to the present theorem it states that the values of the error function of the deviation points alternate in sign. Chebyshev never mentioned this fact, although sometimes the opposite is stated (e. g. in [Nat49] and even in the corresponding article in the Soviet mathematical encyclopedia [Vin77][47]).

Historical commentators disagree. For example, V. L. Goncharov [Gon45] made clear that "indeed in the works of Chebyshev included in the collected works there are no references to an alternation of the sign of the deviation points."[48]

On the other hand, A. A. Gusak wanted to refute this: he cited a situation taken from [Cheb88]. In this text Chebyshev determined a simple joint mechanism where a certain point makes a symmetric movement around an axis (as for example for Watt's parallelogram the center of the connecting-rod).

About the deviation points $\phi_0, \ldots, \phi_3$ of the minimal curve, Chebyshev wrote that this function reaches its maximal value in four points, and in the outer points ϕ_0, ϕ_1 with another sign than in the inner points ϕ_2 and ϕ_3. Gusak shows that because of the property

$$\phi_2 + \frac{\phi_0}{2} = \phi_3 + \frac{\phi_1}{2},$$

proven before in the same text, the signs of the deviation points alternate. So we see that the alternation itself is mentioned by Gusak, not by Chebyshev. So Gusak's considerations only make probable what Goncharov said before: "From that it does not follow that the fact of the alternating signs was not known by him."[49]

It would have been surprising, if he had not remarked it, because a short look at all calculated solutions clearly showed this fact.

But neither of the above-cited corollaries from Theorem 2.1 were important for Chebyshev, since the alternating signs were not necessary for his aim

---

[47] [Vin77, Vol. 5, p. 845]: «Сформулированная теорема была доказана П. Л. Чебышевым в 1854 [...] в более общем виде.»

[48] [Gon45, p. 146]: «Действительно, в работах, вошедших в собрание сочинений Чебышева, нет никаких упоминаний о чередовании знаков отклонения.»

[49] [Gon45, S. 146]: «из чего не следует, конечно, что самый факт чередования знаков не был ему известен.»

to compute solutions of special cases (or as he would have said 'functions least deviating from zero'). The same statement holds for existence or unicity theorems: he found solutions, so that's just pure theory.

At this moment for the first time we want to pay attention to the assumptions Chebyshev set for the functions for which Theorem 2.1 holds. It is conspicuous that Chebyshev only mentioned 'functions'. Necessary assumptions usually were made implicitly. The analyticity of f assumed in the pioneering „Théorie des mécanismes ..." [Cheb54] was no longer demanded, but F also had to be differentiable in x, otherwise the equations (2.30) would not have been valid. Now we know that this assumption is not necessary.

„Sur les questions des minimas ..." unfortunately was the last contribution written by Chebyshev himself which dealt with a general problem of uniform approximation. But a group of papers would follow which dealt with special cases or with problems of least squares approximation.

2.4.8 Chebyshev's Aim

Two years before the monograph „Sur les questions des minima ..." was published Chebyshev gave an overview of possible further results obtained by the methods developed there. In a two-page remark for the bulletin of the St Petersburg Academy [Cheb57] Chebyshev had already summarized the results he regarded as the most important of the main work.

After describing the three cases Chebyshev wrote that "the same approach can be used successfully for many other cases"[50] and there formulated a problem going much farther:

> "We have to find those changes of a given approximated expression for $f(x)$, obtained as usual in the form of a polynomial or a fraction, which the coefficients are subject to, if the maximal error between $x = a - h$ and $x = a + h$ is made as little as possible, h being a sufficiently small number."[51]

Using the methods of best approximation one should find approximations of approximating expressions!

The demanded smallness of h which we know well from section 2.3 lets us presume that the expressions to be approximated are again partial sums of Taylor expansions.

---

[50] [Chebgw2, Bd.S. 148]: «тот же прием может быть выгодно употреблен во многих других случаях.»

[51] [Chebgw2, Bd. 2, S. 148]: «для данного приближенного выражения $f(x)$, выведенного обыкновенными способами в виде многочлена или в виде дроби, найти изменения, которым надо подвергнуть коэффициенты, когда требуется сделать наименьшим предел его погрешностей между $x = a - h$ и $x = a + h$, причем h величина довольно малая.» (Emphasizes by Chebyshev).

2.5 Theory of Orthogonal Polynomials

Chebyshev wrote a series of papers dealing with the approximation of functions in the space $L_2(\rho)$ with several (discrete and continuous) weights. He proved a lot of theorems for the representation of functions using orthogonal polynomials.

2.5.1 On Continuous Fractions

The integration of certain irrational functions led Chebyshev's studies of continuous fractions, which became the ruling means in the monograph discussed in the previous section.

A problem from ballistics by the department of ordnance of the St Petersburg commitee of military sciences[52] prompted him to investigate least square approximation problems due to Gauß.

Without using the term "orthogonality" Chebyshev showed in his work "On Continuous Fractions" [Cheb55/2] which properties orthogonal polynomials satisfy in $L_2(\mu)$ with a discrete measure μ.

The aim of this work was the solution of the following problem:

Let $n+1$ points $x_0, \ldots, x_n$ and values $F(x_0), \ldots, F(x_n)$ be given. We have to determine a polynomial P of degree m $(m < n)$ minimizing the expression

$$\sum_{i=0}^{n} \theta(x_i) \left[F(x_i) - P(x_i) \right]^2 \qquad (2.66)$$

for a weight function θ given in the knots its values $x_0, \ldots, x_n$.

Chebyshev defined

$$f(x) := \prod_{i=0}^{n} (x - x_i)$$

and expanded the expression

$$\frac{f'(x)\theta(x)}{f(x)} \qquad (2.67)$$

into a continuous fraction. We call its jth member ψ_j.

So we obtain a sequence of polynomials

$$\psi_0, \psi_1, \ldots, \psi_m,$$

with

$$\deg \psi_j = j$$

and for which hold

---

[52] Comp. the respective paragraphs about Chebyshev's engagement in the committee of military sciences [Chebgw2, Bd. 5, S. 408 ff.].

$$\sum_{i=0}^{n} \theta(x_i)\psi_j(x_i)\psi_k(x_i) = 0 \quad \text{for} j \neq k \tag{2.68}$$

and

$$\sum_{i=0}^{n} \theta(x_i)\psi_j^2(x_i) = 1. \tag{2.69}$$

These conditions show that the ψ_j build an orthonormal system in $L_2(\mu)$.

Then Chebyshev was able to prove that with the properties (2.68) and (2.69) there holds the following theorem:[53]

Theorem 2.6 *Let $\theta : [a, b] \rightarrow \mathbb{R}^+$ be a discrete measure. Then in $L_2(\theta, [a, b])$ there holds for an orthonormal system of polynomials $\{\psi_0, \ldots, \psi_m\}$ with $\forall j = 1, \ldots, m : \deg \psi_j = j$:*

1. *For all $i \in \{0, \ldots, m\}$ there holds: Among all polynomials of degree i with the same first coefficient, $\|\psi_i\|_{L_2(\theta, [a,b])}$ is minimal.*
2. *The orthogonal projection P with*

$$P(x) := \sum_{i=0}^{m} \frac{\sum_{j=0}^{n} \theta(x_j)\psi_i(x_j)F(x_j)}{\sum_{j=0}^{n} \theta(x_j)\psi_i^2(x_j)} \psi_i(x), \tag{2.70}$$

is the best approximation of F in $L_2(\theta, [a, b])$ according to $\{\psi_0, \ldots, \psi_m\}$.

Proof. [from [Cheb55/2]]

1. Let $i \in \{0, \ldots, m\}$ and V a polynomial of degree i, so that the highest non-vanishing coefficient of V is equal to that of ψ_i. That means V can be represented as

$$V := \sum_{j=0}^{i} A_j \psi_j. \tag{2.71}$$

Because of the above-defined property of the degrees of $\psi_0, \ldots, \psi_m$ we have

$$A_i = 1,$$

because the highest coefficients of V and ψ_i coincide.

We get from (2.71) the following equation for the norm of V

$$\sum_{k=0}^{n} V^2(x_k)\theta(x_k) = \sum_{k=0}^{n} \left[\sum_{j=0}^{i-1} A_j \psi_j(x_k) + \psi_i(x_k) \right]^2 \theta(x_k). \tag{2.72}$$

The right side of equation (2.72) represents a quadratic function of the coefficients $A_0, \ldots, A_{i-1}$. To solve this minimization problem we have at first set the corresponding gradients equal to zero. So we get the equations

---

[53] To keep track we divide Chebyshev's presentation into theorem and proof.

$$2 \sum_{k=0}^{n} \left[\left(\sum_{j=0}^{i-1} A_j \psi_j(x_k) \right) + \psi_i(x_k) \right] \psi_0(x_k) \theta(x_k) = 0$$

$$2 \sum_{k=0}^{n} \left[\left(\sum_{j=0}^{i-1} A_j \psi_j(x_k) \right) + \psi_i(x_k) \right] \psi_1(x_k) \theta(x_k) = 0$$

$$\vdots \qquad \qquad \vdots \ \vdots$$

$$2 \sum_{k=0}^{n} \left[\left(\sum_{j=0}^{i-1} A_j \psi_j(x_k) \right) + \psi_i(x_k) \right] \psi_{i-1}(x_k) \theta(x_k) = 0.$$

Since $\{\psi_0, \ldots, \psi_m\}$ are orthogonal, there remain the terms:

$$2 A_0 \sum_{k=0}^{n} \psi_0^2(x_k) \theta(x_k) = 0$$

$$2 A_1 \sum_{k=0}^{n} \psi_1^2(x_k) \theta(x_k) = 0$$

$$\vdots \qquad \qquad \vdots \ \vdots$$

$$2 A_{i-1} \sum_{k=0}^{n} \psi_{i-1}^2(x_k) \theta(x_k) = 0.$$

From $i \leq m < n$ follows

$$A_1 = A_2 = \cdots = A_{n-1} = 0.$$

So indeed there holds:

$$V \equiv \psi_i.$$

2. The proof of this property is analogous. Now we have values $F(x_0), \ldots, F(x_n)$ and look for the polynomial

$$P(x) := \sum_{j=0}^{m} A_j \psi_j(x),$$

which minimizes the expression (2.66), so here

$$\sum_{i=0}^{n} \left[F(x_i) - \sum_{j=0}^{m} A_j \psi_j(x_i) \right]^2 \theta(x_i). \tag{2.73}$$

To determine this minimum we again differentiate and get the equations

$$-2\sum_{i=0}^{n}\left[F(x_i) - \sum_{j=0}^{m} A_j\psi_j(x_i)\right]\psi_0(x_i)\theta(x_i) = 0$$

$$-2\sum_{i=0}^{n}\left[F(x_i) - \sum_{j=0}^{m} A_j\psi_j(x_i)\right]\psi_1(x_i)\theta(x_i) = 0$$

$$\vdots \qquad\qquad\qquad\qquad \vdots \;\; \vdots$$

$$-2\sum_{i=0}^{n}\left[F(x_i) - \sum_{j=0}^{m} A_j\psi_j(x_i)\right]\psi_m(x_i)\theta(x_i) = 0.$$

Using the orthogonality of $\{\psi_0, \ldots, \psi_m\}$ we are able to simplify:

$$2\sum_{i=0}^{n} F(x_i)\psi_0(x_i)\theta(x_i) - 2A_0\sum_{i=0}^{n}\psi_0^2(x_i)\theta(x_i) = 0$$

$$2\sum_{i=0}^{n} F(x_i)\psi_1(x_i)\theta(x_i) - 2A_0\sum_{i=0}^{n}\psi_1^2(x_i)\theta(x_i) = 0$$

$$\vdots \qquad\qquad\qquad\qquad \vdots \;\; \vdots$$

$$2\sum_{i=0}^{n} F(x_i)\psi_m(x_i)\theta(x_i) - 2A_0\sum_{i=0}^{n}\psi_m^2(x_i)\theta(x_i) = 0.$$

So the sought coefficients are

$$A_0 = \frac{\sum\limits_{i=0}^{n} F(x_i)\psi_0(x_i)\theta(x_i)}{\sum\limits_{i=0}^{n}\psi_0^2(x_i)\theta(x_i)}$$

$$A_1 = \frac{\sum\limits_{i=0}^{n} F(x_i)\psi_1(x_i)\theta(x_i)}{\sum\limits_{i=0}^{n}\psi_1^2(x_i)\theta(x_i)}$$

$$\vdots \;\; \vdots \qquad\qquad \vdots$$

$$A_m = \frac{\sum\limits_{i=0}^{n} F(x_i)\psi_m(x_i)\theta(x_i)}{\sum\limits_{i=0}^{n}\psi_m^2(x_i)\theta(x_i)},$$

and the second part has also been proved. $\qquad\qquad\qquad\qquad\square$

So Chebyshev did recognize the importance of orthogonality. It led him to further investigations about orthogonal function systems.

With the help of Theorem 2.6 we can easily prove a corollary used by Chebyshev (but not explicitly proved):

Corollary 2.7 *Let $\theta : [-1, 1] \rightarrow \mathbb{R}^+$ be a weight function. We define a scalar product in $L_2(\theta, [-1, 1])$ by*

$$< f, g >_{L_2(\theta, [-1,1])} := \int\limits_{-1}^{1} \theta(x) f(x) g(x) \, dx. \qquad (2.74)$$

Now let $\psi_0, \ldots, \psi_m$ be polynomials with

- $\forall j = 1, \ldots, m : \operatorname{grad} \psi_j = j$
- $\{\psi_0, \ldots, \psi_m\}$ *orthonormal with* $< ., . >_{L_2(\theta, [-1,1])}$.

Then we have:

1. *For all $i \in \{0, \ldots, m\}$ there holds: Among all polynomials of degree i having the same highest coefficient $\alpha_i \neq 0$, $\|\psi_i\|_{L_2(\theta, [-1,1])}$ is minimal.*
2. *The orthogonal projection P defined by*

$$P(x) := \sum_{i=0}^{m} \frac{\sum\limits_{j=0}^{m} \int\limits_{-1}^{1} \theta(y) F(y) \psi_j(y) dy}{\int\limits_{-1}^{1} \theta(y) \psi_j^2(y) dy} \psi_i(x) \qquad (2.75)$$

$$= \sum_{i=0}^{m} \frac{\sum\limits_{j=0}^{m} < F, \psi_j >_{L_2(\theta, [-1,1])}}{< \psi_j, \psi_j >_{L_2(\theta, [-1,1])}} \psi_i(x) \qquad (2.76)$$

is the best approximation for F in $L_2(\theta, [-1, 1])$ according to $\{\psi_0, \ldots, \psi_m\}$.

Proof. Theorem 2.6 holds for any interval and discrete weight function θ. It is especially applicable to intervals $[a, b] \subset (-1, 1)$. The Riemannian sums of (2.75) now correspond to expressions of the form (2.70) and converge for any subdivision sequence of the integral (2.75), iff the integrals are taken between a and b instead of -1 and 1. By transition to the limits $a \rightarrow -1$ and $b \rightarrow 1$ we can show this property for the whole interval $[-1, 1]$. $\square$

We could illustrate that only considerations about limits are necessary to solve the continuous case. Later we will explain that Chebyshev usually neglected such reflections, probably because he regarded their results as a matter of course.

The same holds for the questions which properties F should satisfy so that such a theorem will be valid. Of course, Chebyshev did not use expressions like 'scalar product' or 'integrability.'

The method developed by Chebyshev in [Cheb55/2] would later be applied in mathematical statistics under the name regression analysis. Of special importance (and emphasized by Chebyshev himself in [Cheb55/1]) is the simple computation of ϕ_i representing the ith fraction of the expansion of (2.67) and can successively be computed from one to another. So the number of computations was diminished and additionally we got the advantage that the order of the algorithm directly depends on the size of the error and is not determined only by the a priori fixed degree of the interpolation polynomial.

2.5.2 Chebyshev–Fourier Series

Corollary 2.7 gives a series expansion for a function which is L_2-integrable or given by some knots. This expansion depends on the definition of an orthonormal system $\{\psi_0, \ldots, \psi_m\}$ and the scalar product $< ., . >_{L_2(\theta)}$.

Modern theory always calls such series called Fourier series although Fourier discussed only the classical case in [Fou22]. In [Cheb59/2] Chebyshev called these series 'analogous to Fourier series' [54] and showed there some examples:

1. For
$$\theta = \frac{1}{\sqrt{1-x^2}}$$

he got the polynomials $T_1, \ldots, T_m$, later known as Chebyshev polynomials,
2. for
$$\theta = 1$$

he computed the classical Fourier series with the orthonormal system

$$\{1, \cos kx, \sin lx\}$$

and the Legendre-polynomials,[55]
3. for
$$\theta = e^{-x^2}$$

the Hermite polynomials
4. and, finally, for
$$\theta = e^{-x}$$

the Laguerre polynomials.

It is remarkable that Chebyshev investigated both Hermite's and Laguerre's polynomials earlier than they (Hermite and Laguerre introduced the polynomials later named after them as solutions of the respective differential equations in [Her64] and [Lag79]).

---

[54] [Chebgw2, Bd. 2, S. 335]: «аналогичные рядам Фурье.»

[55] also citing his older paper 'On a New Series' [Cheb58], where he first briefly discussed theoretical corollaries from 'On Continuous Fractions'[Cheb55/2].

2.5.3 Theory of Jacobian Polynomials

In his contribution 'About Functions Similar to Legendre's [Cheb70] Chebyshev proved the fact that polynomials of Jacobi type (those which are similar to Legendre's) are orthonormal according to some weight functions.

At first he defined an auxiliary function

$$F(s,x) = \frac{\left(1 + s + \sqrt{1 - 2sx + s^2}\right)^\lambda \left(1 - s + \sqrt{1 - 2sx + s^2}\right)^\mu}{\sqrt{1 - 2sx + s^2}}. \tag{2.77}$$

For $x \in [-1, 1]$ and $s \in (-1, 1)$ this expression is well defined.[56]

He got this idea from a property of Legendre polynomials P_n for which holds:

$$\frac{1}{\sqrt{1 - 2sx + s^2}} = \sum_{n=0}^{\infty} P_n(x)s^n, \quad x \in [-1, 1], \tag{2.78}$$

and the left expression is equal to $F(s,x)$ for $\lambda = \mu = 0$. The right side of equation (2.78) uniformly converges for $s, x \in (-1, 1)$.

The property

$$\int_{-1}^{1} P_n(x)P_m(x) = 0, \tag{2.79}$$

well-known since Legendre's times, can be proved by formally taking

$$\sum_{n=0}^{\infty} \sum_{m=0}^{\infty} s^n t^m \int_{-1}^{1} P_n(x)P_m(x)\, dx = \frac{1}{\sqrt{st}} \log \frac{1 + \sqrt{st}}{1 - \sqrt{st}}.$$

In fact, expanding the right expression into a power series we see that there will be only powers of (st), but no terms $s^m t^n$ with $m \neq n$. Therefore the integral $\int_{-1}^{1} P_n(x)P_m(x)$ for $m \neq n$ will vanish.

Then Chebyshev used the same argument for F.

With lavish calculations using methods from the theory of integration with the help of logarithms he derived himself,[57] he showed that for the integral expression analogous to (2.79) holds:

$$\int_{-1}^{1} \frac{F(s,x)F(t,x)}{(1+x)^\lambda (1-x)^\mu} = \int_{0}^{1} \frac{2^{\lambda+\mu+1}(1 - stz)^\mu}{z^\lambda (1-z)^\mu (1 - stz)^2}\, dz. \tag{2.80}$$

As before it is clear that in the respective series expansion no terms of $s^m t^n$ occur with different degree in s and t.

Using the approach indeed similar to that of Legendre

$$F(s,x) = \sum_{n=0}^{\infty} T_n(x)s^n$$

---

[56] As usual Chebyshev computes formally.
[57] See [Cheb47].

it is easy to show the orthogonality of the Jacobian polynomials T_n in the space[58]

$$L_2\left(\frac{1}{(1+x)^\lambda(1-x)^\mu}\right).$$

2.5.4 Approximation Preserving Monotonicity

The results presented in the previous sections led Chebyshev to an interesting application: the determination of the *monotone polynomial least deviating from zero*. As before he considered polynomials of degree n with a given leading coefficient.

The background of this question again was mechanism theory[Cheb71]. There Chebyshev determined a mechanism, whose movement should be as uniform as possible.[59]

The problem was set in the monograph 'About functions least deviating from zero' [Cheb73]. The mathematical setting is formulated as follows: It is to determine that monotone polynomial n and the form

$$P(x) = x^n + \sum_{i=0}^{n-1} A_i x^i$$

for which $\|P\|_\infty$ is minimal.

2.5.4.1 Preconsiderations

At first Chebyshev showed interesting properties beginning with

$$\|P\|_\infty = |P(1)| = |P(-1)|. \tag{2.81}$$

with P being the polynomial to be determined. Because of its monotonicity the norm is reached at least in one of the end points of the interval. The already well-known shift leads to the sought result, since from $|P(1)| \neq |P(-1)|$ would follow that one of the polynomials

$$\hat{P} = P \pm \frac{P(1)+P(-1)}{2}$$

had a smaller norm than P.

Also it is obvious because of the monotonicity that there holds

$$P(1) = -P(-1).$$

---

[58] To show the validity of these propositions—the integral (2.80) is not defined a priori—one has again to return to considerations about limits, which Chebyshev was definitely not interested in. Compare the arguments regarding Corollary 2.7.

[59] Compare also [Ger54].

Now Chebyshev established a simple, but very interesting property of the solution of the monotone approximation problem.

The main theorem of calculus states that with a constant C there holds:

$$P(x) = \int_{-1}^{x} P'(x)dx + C, \qquad (2.82)$$

whence

$$C = P(-1) = \pm \|P\|_\infty.$$

Of course there is also

$$-C = P(1) = \int_{-1}^{1} P'(x)dx + C,$$

from which follow

$$C = -\frac{1}{2} \int_{-1}^{1} P'(x)dx \qquad (2.83)$$

and

$$\|P\|_\infty = \frac{1}{2} \left| \int_{-1}^{1} P'(x)dx \right|. \qquad (2.84)$$

With these simple considerations Chebyshev observed a connection between this problem and the approximation in the space L_2. Consequently he would use the results from the theory of orthogonal polynomials in the following.

2.5.4.2 Calculation of the Minimal Solution

Since P is monotone the derivative P' has within the interval only roots with even multiplicity. We denote these multiplicities by $2\lambda_1, \ldots, 2\lambda_m$ and by μ_1, μ_2 the multiplicities of the zeros of P' at the ends of the interval. If there are no zeros there, choose respectively $\mu_1 = 0$ or $\mu_2 = 0$.

Now we have

Lemma 2.8 *If P is the non-vanishing monotone polynomial of degree n, for which $\|P\|_\infty$ is minimal, then*

$$\mu_1 + \mu_2 + 2\lambda_1 + \cdots + 2\lambda_m \geq n - 1. \qquad (2.85)$$

Because P is non-vanishing (and trivially not a constant), we have a finite number of zeros of P'. So we have that the number of zeros (counted with multiplicity) is even equal to $n - 1$, so all zeros of P' are real numbers and lie within the interval $[-1, 1]$.

Proof.   We assume the opposite, P' having less than $n - 1$ zeros counted with multiplicity.

Denoting by $\alpha_1, \ldots, \alpha_m$ the zeros of P' within $(-1, 1)$ we have that the function

$$Q(x) = \frac{P'(x)}{(x-1)^{\mu_1}(x+1)^{\mu_2} \prod\limits_{i=1}^{m}(x-\alpha_i)^{2\lambda_i}} \tag{2.86}$$

is non-constant and has by definition no zero in the interval $[-1, 1]$. Without losing generality we assume that Q is positive. Set

$$L_0 := \min_{x \in [-1,1]} |Q(x)|.$$

So $Q - L_0$ remains non-negative. Hence we have for all $x \in [-1, 1]$:

$$Q(x) > Q(x) - L_0 \geq 0, \quad \text{and so}$$

$$P'(x) > P'(x) - L_0(x-1)^{\mu_1}(x+1)^{\mu_2} \prod_{i=1}^{m}(x-\alpha_i)^{2\lambda_i} \geq 0. \tag{2.87}$$

Because the integral is a monotone operator we have

$$\int_{-1}^{1} P'(x)\, dx > \int_{-1}^{1} P'(x) - L_0(x-1)^{\mu_1}(x+1)^{\mu_2} \prod_{i=1}^{m}(x-\alpha_i)^{2\lambda_i}\, dx.$$

By assumption the degree of the function

$$L_0(x-1)^{\mu_1}(x+1)^{\mu_2} \prod_{i=1}^{m}(x-\alpha_i)^{2\lambda_i}$$

is less than $n - 1$, so there holds with (2.87) and (2.84):

$$2\|P\|_\infty > \int_{-1}^{1} P'(x) - L_0(x-1)^{\mu_1}(x+1)^{\mu_2} \prod_{i=1}^{m}(x-\alpha_i)^{2\lambda_i}\, dx > 0.$$

This shows that P cannot be the monotone polynomial of least deviation from zero. $\qquad\square$

With this result we get a representation of the derivative

$$P'(x) = C(x-1)^{\mu_1}(x+1)^{\mu_2} \prod_{i=1}^{m}(x-\alpha_i)^{2\lambda_i} \tag{2.88}$$

with a constant C.

Since P has the form

$$P(x) = x^n + \sum_{i=0}^{n-1} x^i,$$

it follows that

$$C = n.$$

Setting

$$\mu_i = 2\kappa_i + \nu_i, \quad i = 1, 2$$

with

$$\nu_i \in \{0, 1\}$$

we can write

$$P'(x) = n(x-1)^{\nu_1}(x+1)^{\nu_2}\left[(x-1)^{\kappa_1}(x+1)^{\kappa_2}\prod_{i=1}^{m}(x-\alpha_i)^{\lambda_i}\right]^2, \quad (2.89)$$

or as an abbreviation

$$P'(x) = n(x-1)^{\nu_1}(x+1)^{\nu_2}U^2(x) \quad (2.90)$$

with a polynomial U of degree

$$l := \kappa_1 + \kappa_2 + \sum_{i=1}^{m}\lambda_i.$$

To minimize the integral

$$\|P\|_\infty = \frac{1}{2}\left|\int_{-1}^{1}P'(x)\,dx\right| = \frac{n}{2}\left|\int_{-1}^{1}(x-1)^{\nu_1}(x+1)^{\nu_2}U^2(x)\,dx\right|, \quad (2.91)$$

we can fall back on some results from [Cheb70] cited in previous paragraphs. For the polynomial of degree l and leading coefficient 1, minimizing expression (2.91) is just the polynomial T_l of the series expansion

$$F(s, x) = \sum_{l=0}^{\infty}T_l(x)s^l,$$

where

$$F(s, x) = \frac{\left(1+s+\sqrt{1-2sx+s^2}\right)^{-\nu_1}\left(1-s+\sqrt{1-2sx+s^2}\right)^{-\nu_2}}{\sqrt{1-2sx+s^2}}. \quad (2.92)$$

With this compare the considerations directly after equation (2.77) on page 57.

So we can set in our special case

$$U = KT_l.$$

Let K_l be the leading coefficient of T_l. Therefore we have with (2.90)

$$C = \frac{1}{K_l} \quad \text{and}$$

$$U = \frac{1}{K_l} T_l. \tag{2.93}$$

To determine the function T_l we firstly define its coefficients by

$$T_l =: K_0 + K_1 x + \cdots + K_l x^l.$$

Then we get the series expansion of (2.92) as

$$F(s, x) = \frac{\left(1 + s + \sqrt{1 - 2sx + s^2}\right)^{-\nu_1} \left(1 - s + \sqrt{1 - 2sx + s^2}\right)^{-\nu_2}}{\sqrt{1 - 2sx + s^2}}$$

$$= \sum_{l=0}^{\infty} \left(\sum_{i=0}^{l} K_i x^i \right) s^l. \tag{2.94}$$

To simplify this expression we take a more careful look at the expansion round.[60] $s_0 = 0$.

With this we set

$$u(s, x) := 1 + s + \sqrt{1 - 2sx + s^2}$$
$$v(s, x) := 1 - s + \sqrt{1 - 2sx + s^2}$$
$$f(s, x) := 1 - 2sx + s^2$$

and compute

$$\frac{\partial u}{\partial s}(s, x) = 1 - \frac{x - s}{\sqrt{1 - 2sx + s^2}} \qquad u(0, x) = 2 \qquad \frac{\partial u}{\partial s}(0, x) = 1 - x \quad (2.95)$$

$$\frac{\partial v}{\partial s}(s, x) = -1 - \frac{x - s}{\sqrt{1 - 2sx + s^2}} \qquad v(0, x) = 2 \qquad \frac{\partial v}{\partial s}(0, x) = -1 - x$$

$$\tag{2.96}$$

$$\frac{\partial f}{\partial s}(s, x) = 2s - 2x \qquad f(0, x) = 1 \qquad \frac{\partial f}{\partial s}(0, x) = -2x. \quad (2.97)$$

Then

$$F = u^{-\nu_1} v^{-\nu_2} f^{-\frac{1}{2}},$$

$$\frac{\partial F}{\partial s} = -\nu_1 u^{-\nu_1 - 1} \frac{\partial u}{\partial s} v^{-\nu_2} f^{-\frac{1}{2}} + -\nu_2 v^{-\nu_2 - 1} \frac{\partial v}{\partial s} f^{-\frac{1}{2}} - -\frac{1}{2} f^{-\frac{3}{2}} \frac{\partial f}{\partial s} u^{-\nu_1} v^{-\nu_2}$$

[60] Chebyshev had this in mind writing "set $s = 0$" ([Chebgw3, S. 593]: «полагая $s = 0$.»)

and

$$\frac{\partial F}{\partial s}(0,x) = \left((\nu_1 + \nu_2)2^{-(\nu_1+\nu_2)-1} + 2^{-(\nu_1+\nu_2)}\right)x - (\nu_1 + \nu_2)2^{-(\nu_1+\nu_2)-1}.$$

To calculate the minimal solution U, only the coefficient of the highest power is of importance. It can be determined by a significant simplification of F.

Consider

$$u_2(s,x) := 1 + \sqrt{1 - 2sx}$$
$$f_2(s,x) := 1 - 2sx$$

and calculate

$$\frac{\partial u_2}{\partial s}(s,x) = -\frac{x}{\sqrt{1 - 2sx}} \qquad u_2(0,x) = 2 \qquad \frac{\partial u_2}{\partial s}(0,x) = -x \qquad (2.98)$$

$$\frac{\partial f_2}{\partial s}(s,x) = -2x \qquad f_2(0,x) = 1 \qquad \frac{\partial f_2}{\partial s}(0,x) = -2x. \qquad (2.99)$$

Now set

$$F_2 := u_2^{-(\nu_1+\nu_2)} f_2^{-\frac{1}{2}}$$

and compute

$$\frac{\partial F_2}{\partial s} = -(\nu_1 + \nu_2)u_2^{-(\nu_1+\nu_2)-1}\frac{\partial u}{\partial s}f_2^{-\frac{1}{2}} - \frac{1}{2}f_2^{-\frac{3}{2}}\frac{\partial f}{\partial s}u_2^{-(\nu_1+\nu_2)}$$

and

$$\frac{\partial F_2}{\partial s}(0,x) = \left((\nu_1 + \nu_2)2^{-(\nu_1+\nu_2)-1} + 2^{-(\nu_1+\nu_2)}\right)x.$$

This construction shows that the highest coefficients of F and F_2 do not differ from each other continuing the expansion round $s_0 = 0$.

So the sought coefficients K_l in the series expansion (2.94) can also be taken from the expansion

$$F_2(s,x) = \sum_{l=0}^{\infty} K_l x^l s^l.$$

The numbers ν_1 and ν_2 are remainders of divisions by 2 and so are equal to 0 or 1, that is, the sum $\nu_1 + \nu_2$ is equal to $0, 1$ or 2.

Case 1: $\nu_1 + \nu_2 = 0$

Then

$$F_2(s,x) = (1 - 2sx)^{-\frac{1}{2}}$$
$$\frac{\partial^l F_2}{\partial s^l}(0,x) = 1 \cdot 3 \cdots (2l-1)x^l = \frac{(2l-1)!!}{2(l+2)}x^l$$
$$K_l = \frac{(2l-1)!!}{l!}.$$

Case 2: $\nu_1 + \nu_2 = 1$
Then

$$F_2(s, x) = (1 + \sqrt{1 - 2sx})^{-1}(1 - 2sx)^{-\frac{1}{2}}$$

$$\frac{\partial^l F_2}{\partial s^l}(0, x) = \frac{1}{2(l+1)} 1 \cdot 3 \cdots (2l + 1)x^l = (2l - 1)!! \, x^l$$

$$K_l = \frac{(2l+1)!!}{2(l+1)!}.$$

Case 3: $\nu_1 + \nu_2 = 2$
Then

$$F_2(s, x) = (1 + \sqrt{1 - 2sx})^{-2}(1 - 2sx)^{-\frac{1}{2}}$$

$$\frac{\partial^l F_2}{\partial s^l}(0, x) = \frac{1}{2(l+2)} 1 \cdot 3 \cdots (2l + 1)x^l = (2l - 1)!! \, x^l$$

$$K_l = \frac{(2l+1)!!(l+1)}{2(l+2)!}.$$

Now it is easy to determine ν_1 and ν_2 by looking at the four possible of even or odd n and a monotone decreasing or increasing P and analysing (2.89).

So l can be determined:

$$l = \frac{n - \nu_1 - \nu_2 - 1}{2},$$

and by the series expansion of $F(s, x)$ (2.92) one can get T_l. So we achieve

$$P'(x) = \frac{n}{K_l^2}(x - 1)^{\nu_1}(x + 1)^{\nu_2}T_l^2,$$

and from the equations (2.82) and (2.83) there finally follows

$$P(x) = \frac{n}{K_l^2}\left(\int_{-1}^{x}(x - 1)^{\nu_1}(x + 1)^{\nu_2}T_l^2 \, dx - \frac{1}{2}\int_{-1}^{1}(x - 1)^{\nu_1}(x + 1)^{\nu_2}T_l^2 \, dx\right).$$

2.5.4.3 Calculation of the Minimal Deviation

Equation (2.91) gives the value of the minimal deviation L. So it is

$$L = \frac{1}{2}\left|\int_{-1}^{1} P'(x) \, dx\right|.$$

With (2.90) there follows

$$L = \frac{n}{2} \left| \int_{-1}^{1} (x-1)^{\nu_1}(x+1)^{\nu_2} U^2(x)\, dx \right|.$$

The integrand has degree $n-1$. Since $\deg U = l$ we finally have from (2.93)

$$L = \frac{2l + \nu_1 + \nu_2 + 1}{2K_l^2} \left| \int_{-1}^{1} (x-1)^{\nu_1}(x+1)^{\nu_2} U^2(x)\, dx \right|.$$

The integral

$$I := \int_{-1}^{1} (x-1)^{\nu_1}(x+1)^{\nu_2} U^2(x)\, dx$$

can be calculated by results from [Cheb70]. Setting for λ and μ the values $-\nu_1$ and $-\nu_2$, respectively, then from equation (2.80) there follows in this case

$$\int_{-1}^{1} \frac{\sum_{m=0}^{\infty} T_m s^m \sum_{k=0}^{\infty} T_k t^k}{(1-x)^{-\nu_1}(1+x)^{-\nu_2}}\, dx = \int_{0}^{1} \frac{2^{-\nu_1 - \nu_2 + 1}(1 - stz)^{-\nu_1}}{z^{-\nu_2}(1-z)^{-\nu_1}(1 - stz)^2}\, dz =: G(s,t).$$

We see that I is the coefficient of the lth power in the series expansion of $G(s,t)$.

Now we have to look at the three well-known cases again.

Case 1: $\nu_1 + \nu_2 = 0$

Then

$$G(s,t) = \frac{1}{\sqrt{st}} \log \frac{1 + \sqrt{st}}{1 - \sqrt{st}}$$

$$I = \frac{2}{2l+1}$$

$$L = \left(\frac{l!}{(2l-1)!!} \right)^2 = \left(\frac{\left(\frac{n-1}{2}\right)!}{(n-2)!!} \right)^2.$$

Case 2: $\nu_1 + \nu_2 = 1$

Now

$$G(s,t) = -\frac{1}{2st} \log(1 - st)$$

$$I = \frac{1}{2(l+1)}$$

$$L = 2 \left(\frac{(l+1)!}{(2l+1)!!} \right)^2 = 2 \left(\frac{\left(\frac{n}{2}\right)!}{(n-1)!!} \right)^2.$$

Case 3: $\nu_1 + \nu_2 = 2$

Here

$$G(s,t) = \frac{1}{2}\left[\frac{1}{2(st)^{-\frac{3}{2}}}\log\frac{1+\sqrt{st}}{1-\sqrt{st}} - \log\frac{1-st}{s^2t^2}\right]$$

$$I = \frac{l+1}{2(l+2)(2l+3)}$$

$$L = \frac{l+2}{l+1}\left(\frac{(l+1)!}{(2l+1)!!}\right)^2 = \frac{n+1}{n-1}\left(\frac{(\frac{n-1}{2})!}{(n-2)!!}\right)^2.$$

Chebyshev used these values to give statements about the monotonicity of polynomials of the form

$$x^n + \sum_{i=0}^{n-1} a_i x^i.$$

Finally he can show the improvements of approximations of zeros of certain equations mentioned in section 2.4.5.

Indeed, after similar considerations there is

Corollary 2.9 [Chebgw3, S. 606] *Let p be a polynomial of degree n on* $[-h,h]$ *and consider the following form:*

$$p(x) := x^n \sum_{i=1}^{n-1} a_i x^i.$$

Then there holds for all $t \in [-h,h]$:

Between the points t and $t - 4\operatorname{sign}\frac{f(t)}{f'(t)}\sqrt[n]{\frac{f(t)}{2(n-1)\pi}}$ *there is a point, where f and f' have a different sign.*

And furthermore:

Theorem 2.10 [Chebgw3, S. 608] *The equation*

$$x^{2l+1} + \sum_{i=0}^{l-1} a_i x^{2i} = 0 \tag{2.100}$$

has at least one zero in the interval

$$\left[-2\sqrt[2l+1]{\frac{a_0}{4l\pi}}, 2\sqrt[2l+1]{\frac{a_0}{4l\pi}}\right].$$

This result was improved later by A. A. Markov in [MarA03].

2.6 Other Contributions of P. L. Chebyshev

Besides the above-described works there are some other papers of Chebyshev dealing with aspects similar to approximation theory. They affect the theory of mechanisms, investigations about solutions of algebraic equations, applications to geometry and quadrature formulae.

This enumeration already makes clear his interest in the application of mathematical results.

2.6.1 Theory of Mechanisms

As we have already seen, the theory of mechanisms had been Chebyshev's trigger to deal with the theory of best approximation.

It was born from the search for a good mechanism and was based on his own theory, developed in his two famous monographs, where he at first only made tentative attempts to special solutions [Cheb54], until he finally fixed surprisingly general theoretical foundations [Cheb59].

It is quite a bit suprising that Chebyshev's first results on mechanism theory were published only in 1861 [Cheb61], although he had already announced a second part of the 'Theory of mechanisms ...' in 1853. It seems that he realized that his theoretical results had not been sufficiently good.[61]

Chebyshev's work on mechanism theory on the one hand deals with improvements of well-known mechanisms like Watt's parallelogram, on the other hand with the construction of mechanisms carrying out the same movements with fewer components.

So he managed to build a mechanism realizing the transformation from a circular into a linear motion, so replacing Watt's parallelogram, with only three joints [Cheb88] and a simplification of Galloway's mechanism which transmits a rotation doubling the radius (ibidem).

All the methods he used there were based on results from his investigations about approximation theory, since all improvements spread over all points of the curves describing the respective motions.

Chebyshev not only calculated but also physically made new mechanical constructions, joint mechanisms of all kinds, a machine simulating human running motions and a calculating machine. Artobolevski and Levitski [ArLe45, S. 107-109] count 41 inventions constructed by Chebyshev himself.

His constructions can be visited in museums of St Petersburg; his calculator is even exhibited in the Paris Consérvatoire des arts et métiers.[62]

---

[61] With this compare the explanations of see S. N. Bernstein in [Bern47, p. 41].

[62] There are several detailed comments about Chebyshev's work on mechanism theory, compare e. g. [Del00], [ArLe45] and [ArLe55/1].

2.6.2 Geodesy and Cartography

Three of Chebyshev's contributions are of isolated character and devoted to applications to geoscience.

The short paper 'A Rule for an Approximative Determination of Distances on the Earth's Surface' [Cheb69] applied Poncelet's approximation formulae to an approximative calculation of the geodetic line between two points given by their corresponding longitudes and latitudes.

To do this Chebyshev firstly used a variant of Pythagoras' theorem to compute the arclength of the grand circle between the two points to get the premisses to apply Poncelet's formulae.

The text 'About the Construction of Geographic Maps' [Cheb56/1] and the speech 'Drawing Geographic Maps,' [Cheb56/2] which we will talk about later for other reasons, deal with a problem from cartography, to be precise the search for a conformal map projection (preserving angles), where the error of the scale logarithm will be as small as possible.

With Gauß' 'Theorema egregium' there is no conformal projection which preserves the scale, therefore the problem is well posed.

Without giving a proof Chebyshev stated for a solution of this question that the scale should remain constant on the border of the map.[63]

This question led to an approximation with two variables. A proof of the correctness of Chebyshev's proposition was made much later by D. A. Grave [Gra96] and [Gra11].

2.6.3 Approximated Quadrature Formulae

In two of his last papers Chebyshev discussed the approximative representation of quadrature formulae, mainly for the approximative determination of elliptic integrals.

In his article 'About Approximations of a Square Root [...] by Simple Fractions' [Cheb89] Chebyshev solved the problem to approximate the function $\frac{1}{\sqrt{x}}$ by expressions of the form

$$A + \sum_{i=1}^{n} \frac{B_i}{C_i + x} \tag{2.101}$$

in the sense of best approximation. With this he got a good approximation also for certain elliptic integrals.

For example, Chebyshev investigated the integral

$$\int \frac{\tan^{p-1} x}{\sqrt{1 - \lambda^2 \sin^2 x}} \, dx.$$

---

[63] This follows from Dirichlet's principle—it is not known whether Dirichlet, who Chebyshev met before 1856, influenced the solution of this problem.

The second contribution 'About Polynomials...' [Cheb93] was devoted to the approximation of expressions of type (2.101) by polynomial expressions.

Here he used the approximation

$$\frac{1}{H-x} \sim \frac{1}{T_n(\frac{H}{h})} \times \left(T_n(\frac{x}{h}) - T_n(\frac{H}{h}) \right),$$

on the interval $[-h, h]$, where $h < H$.

Again Chebyshev discussed uniform approximations controlling the error on the whole interval.

These considerations led to the approximation of integral expressions like

$$\int \frac{f(x)}{H-x} \, dx$$

by linear combinations of more simple expressions, namely

$$\int x^k f(x) \, dx.$$

2.7 Chebyshev–Euler of the 18th Century?

The contributions discussed here only cover a small part of Chebyshev's whole work. Vasilev [Vas00] counted 77 published papers all written by Chebyshev without any co-authors which spread over many subjects, besides approximation and mechanism theory also number theory, the studies of elliptic integrals (often named by Chebyshev as the integration of irrational differentials) and probability theory.

Especially his early work was devoted to number theory, for example his doctorate thesis about the theory of congruences (published as [Cheb49]), which would be translated in several languages, and his fame arose through this subject because of his proof of the prime number theorem [Cheb48].

As we could see elliptic integrals played a very important methodological role in his work. He first investigated them in his dissertation pro venia legendi [Cheb47].

He gave theoretical foundations to probability theory (in his master thesis [Cheb45]) and opened it to practical applications beyond the casino.

These wider branches of his interest are supplemented by some contributions to special problems, as we already observed before. Surely his strangest paper was his talk before the Association Française pour l'avancement des sciences devoted to the subject 'About the Cut of Clothes'[Cheb78]. But the curiosity of his work about mechanisms was not less remarkable.

In spite of the diversity of Chebyshev's work all of its parts seem to be connected with each other. During his life, full of hard scientific work, Chebyshev often returned to results which he had developed earlier and applied them to get new ones.

The approximation theory is a good example to make this clear. Chebyshev used results from his dissertation pro venia legendi [Cheb47] to calculate T_n in his work ,,Théorie des mécanismes...'' [Cheb54] in 1854 , to prove the orthogonality of Jacobian polynomials in [Cheb70] in 1870 and in connection with the computations of approximated quadrature formulae (compare section 2.6.3). Except for a small remark and a condition for the existence of closed formulae for elliptic integrals by Abel, which had not been helpful for practical purposes, there had not been any results of other authors on this subject before.

Similar facts hold for the theory of continuous fractions, which Chebyshev had discovered as a helpful instrument in 1855 (,,About Continuous Fractions'' [Cheb55/2]), laid out the foundations of the theory of orthogonal polynomials and generalized Fourier series and let them work in an unbelievably virtuoso manner in ,,Sur les questions de minima ...'' [Cheb59] to solve the third case about the best rational approximation.

So Chebyshev's work was very stringent and it is still astonishing how many applications it covered and how straight it went from pure mathematics to concrete solutions of practical problems.

This had a programmatic character in Chebyshev's scientific life, as he himself emphasized in his speech 'Drawing Geographical Maps' [Cheb56/2]:

> "The congregation of theory and practice gives the best results, and it is not only practice which gains a benefit from this; science itself is developing under the influence of practice.[...] If the theory gets much from new applications of an old method [...], then it gains even more from the discovery of new methods, and in this case science finds itself a true leader in the practice."[64]

It has to be emphasized, however, that the concept 'practice' has to include applications within mathematics themselves—otherwise it would not be possible to understand Chebyshev's early work regarding the theory of numbers. Without any doubt, for Chebyshev the aim of all mathematical effort should be a (at least approximative) solution of a practical problem.

According to the memories of Andrey Andreevich Markov junior, the son of Chebyshev's pupil Andrey Andreevich Markov, Chebyshev a bit jokingly regards this as the result of a historical development. He is cited:

---

[64] «Сближение теории с практикой дает самые благотворные результаты, и не одна только практика от этого выигрывает; сами науки развиваются под влиянием ее; [...] Если теория много выигрывает от новых приложений старой меюды [...], то она еще более приобретает открытием новых меюд, и в этом случае наука находит себе верного руководителя в практике.»

"In ancient times mathematical problems were posed by gods, in the middle ages by czars and kings, and in our times - by need."[65]

Despite the extent of his work it is conspicuous that Chebyshev had nearly no real colleagues he moved around alone in all his favourite subjects, not only in the approximation theory, where his ideas had almost no forerunner.

Commenting on this his pupil K. A. Posse wrote [Pos04]: "We have to remark that Chebyshev himself preferred working independently to reading others' work. After substantial studies of the work of famous mathematicians— of Euler, Lagrange, Gauß, Abel and others—Chebyshev did not attach much importance to reading modern mathematical literature, emphasizing that unnecessary studies of other authors' papers would have a harmful effect on the independence of one's own work."[66]

The versatility and novelty of his ideas, all being classical in style and often elementary, caused the next generation to regard him with Lobachevski as the most significant mathematician of the 19th century. For example in 1945 S. N. Bernstein wrote in a contribution for the popular-science journal 'nature' (природа, [Bern45/1]) - not neglecting the patriotic undertone, forced at that time:

"However remarkable might be the achievements of modern Russian mathematics growing from day to day, however sparkling they may be in the future, there are two immortal names in the history of our sciences which occupy an *especially*[67] honourable place in it forever. They are—Lobachevski and Chebyshev—two poles of mathematical thinking, who firstly and nearly simultaneously discovered the special power, originality and versatility of the Russian genius for the world[68] [...]"

---

[65] Cited after [Gro87, S. 30]: «[...] в древние времена математические задачи ставились богами, в средние века - царями и королями, а в наши времена - нуждой.»

[66] [Chebgw2, Bd. 5, S. 7]: «Надо заметить, что сам Чебышев более любил самостоятельные исследования, чем изучение трудов других математиков, особенно современных. Глубоко изучив творения великих математиков - Эйлера, Лангранжа, Гаусса, Абеля и других, Чебышев не придавал особого значения чтению текущей математической литературы, утверждая, что излишнее усердие в изучении чужих трудов должно неблагоприятно отражаться на самостоятельности собственных работ.»

[67] The emphasis was made by Bernstein in the contribution [Bern47], which is equal in words.

[68] [Bern45/1, S. 78]: «Как бы значительны ни были достижения современной русской математики, растущие с каждым днём, как бы блестящи ни были её будущие успехи, есть два бессмертных имени в истории нашей науки, которые всегда будут занимать в ней особо почётное место. Это - Лобачевский и Чебышев - два полюса математической мысли, которые впервые открыли миру исключительную мощь, оригинальность и многогранность русской математической гения, почти одновременно [...]»

B. N. Delone, a mathematician in the subject of number theory and corresponding member of the Soviet academy of sciences of that time,[69] tried to surpass him in the same year during a speech to the 220th anniversary of the Russian-Soviet academy[Del45]:

"The works of Chebyshev are unusually versatile [...]. In this sense he resembles the great classics of our time, Euler and Lagrange."[70]

In one point, significant at least from today's point of view, Chebyshev's and Euler's works clearly differ.[71] Euler also got an outstanding significance by his new foundations of modern analysis in the „Introductio in analysin infinitorum" [Eul47].

Especially Euler's concept of a function was epoch-making because it could abstract from the identity between function and graph which was due to Leibniz [Eul47, § 4]:

"A function of a variable quantity is an analytic expression, which is in some way composed by that variable quantity, numbers and constant quantities."[72]

After that he divided functions into two main classes: algebraic and transcendent,[73] and this terminology holds until now. It is remarkable that later during the peak of the discussion about the solutions of the wave-equation[74] he widened the concept of a function and was led to a very modern version

---

[69] By the way, he was not the author of contribution [Del00], N. B. Delone (1856–1931), but his son.

[70] [Del45, S. 4]: «Работы Чебышева необыкновенно разнообразны [...] В этом Чебышев подобен великим классикам нашей науки, Эйлеру и Лангранжу.»

[71] Besides this we have to remark that the number of Euler's papers enormously exceeds that of Chebyshev. His complete collected works are still being edited and already contain fifty volumes.

[72] [Eul47, § 4]: „Functio quantitatis variabilis est expressio analytica quomodocunque composita ex illa quantitate variabili et numeris seu quantitatibus constantibus."

[73] [Eul47, § 7]: „Functiones dividuntur in algebraicas et transcendentes [...]"

[74] The solutions of the wave-equation

$$\frac{\partial^2 y}{\partial t^2} = a^2 \frac{\partial^2 y}{\partial x^2}, \quad x \in [0, 1]$$

caused a very intensive discussion about the concept of a function. The solutions found by d'Alembert in 1748 were not only defined on the interval $[0, 1]$, and of course one had not had any information about their properties beyond this interval. He himself claimed a principle of continuity, which meant that only those which were solutions for this equation on all intervals were suitable. Euler rejected this a posteriori condition and its principle of continuity and so set a milestone for the further formation of mathematical theory (for a more detailed discussion compare [Vol87, p. 157 ff.]).

"When quantities depend on other quantities so that these change themselves, when those also do, then the first are called functions of the last. This definition is of a very general nature and covers all methods which help to define one quantity by others. So, if x denotes a variable quantity, then all quantities depending on x or being defined by x are called functions of x."[75]

The above-cited words of Posse suggest that Chebyshev knew Euler's work . Besides this Chebyshev studied Euler's papers since his time at Moscow university, because they had been recommended to the students with those of Cauchy and Lacroix.[76] Initiated by Euler's great grandson, the academician P. N. Fuss, he later supported Bunyakovski with the edition of some rediscovered contributions of Euler to the theory of numbers,[77] and so we can assume that he indeed had a good overview of Euler's work and knew the above-mentioned concepts of a function.

By the same reasoning, he probably knew about Cauchy's attempts to define the continuity of a function.

But reading Chebyshev's work regarding this aspect we see that Chebyshev's functions had always implicitly more properties to satisfy than the more general concept of Euler demanded, although Chebyshev only called them 'functions'. If they should satisfy more properties, they do. Shortly (and slightly polemically) spoken, for Chebyshev a function was something which had as many derivatives as one needed.

But it is also possible that he firstly thought that a continuous function is at least piecewise differentiable. This had been a usual opinion at that time as many attempts to prove it have. Only with Weierstraß' discovery of a continuous function that is nowhere differentiable, was this question negatively answered. We can assume that Chebyshev knew about new developments within the Weierstraß school, since two of his pupils, A. V. Bessel' in 1862 and D. F. Selivanov in 1880[78] studied at the Berlin university through Chebyshev's initiative.

We have already seen that it was not very important for Chebyshev to check all theoretical implications of his theorems (compare the discussion about the alternation theorem), since he was firstly interested in solutions of practical problems.

---

[75] [Eul55, S. 4]: „Quae autem quantitatis hoc modo ab aliis pendent, ut his mutatis etiam ipsae mutationes subeant, eae harum functiones appellari solent; quae denominatio latissime patet atque omnes modos, quibus una quantitas per alias determinari potest, in se complectitur. Si igitur x denotet quantitatem variabilem, omnes quantitates, quae utcunque ab x pendent seu per eam determinantur, eius functiones vocantur [...]"

[76] Comp. [Pru76, p. 40].

[77] Comp. [Prú76, p. 70].

[78] See [Erm97, S. 2].

There are only a few sources which present his opinions about questions of the theoretical foundations of mathematics which were actual at that time. He did not discuss this in his own research papers and at the university he did not lecture on basics where he should express his thoughts about the foundations of mathematics.[79]

Therefore we can only cite one source where Chebyshev himself talks about these problems, the notes of one listener to his lecture about probability theory (1876/77), which contain many details and especially statements that Chebyshev made in passing to illustrate the results he presented. It was justified that the re-discovery of these notes by N. S. Ermolaeva [Erm87] caused a certain stir.

An interesting remark about mathematicians themselves is:

"I divide mathematicians into two categories: the ones who deal with mathematics to solve new problems from nature and whose results are clear and the others who love mathematics as a subject to philosophize about. I think that the second ones should not be called mathematicians, and who falls on this way won't be a mathematician. According to the question about the statute of the university I therefore clearly said that at the faculty of mathematics[80] lectures about philosophy are not desirable."[81]

We can only guess who he had concretely in mind (later Lyapunov would become clearer), but a little later talking about 'infinitely small quantities' he reveals himself:

"You have already recognized that sometimes I come down to philosophizing, but then it is not about the subject, not about what a quantity, a space are like, but about methods. I do not think about the origins, but about what should be in mind for the solution of new problems. For this all kinds of considerations are useful. But philosophizing about what an infinitely small quantity is, does not lead to

---

[79] Compare the surveys in [Chebgw2, Bd. 5] and the lecture timetables of St Petersburg University [VorSPb], which are available from 1869.

[80] Already in 1861 there was a discussion about the introduction of optional subjects like logic and history of philosophy, since there was the opinion that from these subjects students could also gain a profit for their own scientific work. October 29, 1876 there was a second meeting on this subject (shortly before the cited lecture) and Chebyshev spoke against this suggestion. (See [Erm97])

[81] Cited after [Erm97]: «Я разделяю людей-математиков на две категории: одни, занимающиеся математикою для решения новых вопросов из природы, результаты которых осязательны, или вопросов геометрии и прочих; а другие, которые любят математику как предмет философствования. По-моему, вторые не должны считаться и математиками, и кто попадет на эту дорогу, тот не будет математиком. Поэтому при вопросе о рассмотрении университетского устава я прямо высказался, что преподавание философии на математическом факультете не желательно.»

anything. Here we have one of two alternatives: Either we go via philosophizing to the point that conclusions using infinitely small quantities were not strict—and then we would have to reject the infinitely small—or we come to the fact to prove the correctness of those conclusions. The experience shows that all people who dealt with similar questions explained this only for themselves and did not add anything for the solution of new problems. In these cases I do not recommend you to philosophize."[82]

Clearly someone who talked about the validity of infinitely small quantities in such a manner would also doubt the sense of the exact definition of continuity and the tightly connected question—for which functions his theorems hold. He might object that all the theorems are valid for the respective functions met in practical cases.

Of course it became clear that Weierstraß' approximation theorem, proved in 1885, left no trace in Chebyshev's work. Presumably he was not surprised by the result because in some way it confirmed his position to investigate only functions satisfying certain desired properties—finally all continuous functions are limits of good-natured analytic ones, even the degenerate non-differentiable continuous ones.

So a large part of the work of Western European mathematicians passed by their Russian colleagues without having great influence on them. We will see in analysing the work of Chebyshev's pupils that it would take another generation until the discussion about the foundations of mathematics fell on fertile ground and was used in approximation theory.

---

[82] Cited after [Erm97]: «Вы заметили, что на лекциях я иногда пускаюсь в философствование, но это не о предмете, не о том, что такое величина, пространство, а относительно методов. Думаю не о том, что в основе, а о том, что должно иметь в виду при решении новых вопросов. При этом всякого рода соображения бывают полезны. А философсвствование о том, что такое бесконечно малая величина, ни к чему не ведет. Тут может быть одно из двух: или чрез философствование дойти до того, что выводы с бесконечно малыми величинами нестроги, и тогда пришлось бы отказаться от бесконечно малых, - или на деле подтвердить строгость этих выводов. Опыт показывает, что все люди, занимающиеся подобными вопросами, все только для себя разъясняли дело и ничего не прибавили к тому, чтобы решить новые вопросы. В таких случаях советую вам не философствовать.»

3

The Saint Petersburg Mathematical School

Pafnuti Lvovich Chebyshev's significance has a second basis because of his role as founder of a mathematical school. Already before him there were famous Russian mathematicians like Lobachevski or at least mathematicians who worked in Russia for a long time like Euler, but for the time being their work hardly had an influence on the scientific surroundings because there had not been any pupils who continued working on their ideas.

So we meet other circumstances with Chebyshev. His results defined new scientific directions, his pedagogic jobs, especially at the Imperial St Petersburg University from 1847 until 1882, caused interest in these subjects.

We may name as his 'direct successors'[1] Aleksandr Nikolaevich Korkin[2] (1837–1908), A. V. Bessel (1839–1870), Julian Karol Sochocki[3] (1842–1927), Matvej Aleksandrovich Tikhomandritski[4] (1844–1921), Egor Ivanovich Zolotarev[5] (1847–1878), Konstantin Aleksandrovich Posse[6] (1847–1928), Nikolay Yakovlevich Sonin[7] (1849–1915), Aleksandr Vasilevich Vasilev[8] (1853–1929), Ivan Lvovich Ptashitski[9] (1854–1912), Dmitri Fedorovich Selivanov[10]

---

[1] We want to name as 'direct successors' those mathematicians Chebyshev himself had an influence on as a teacher or a promoter, so not only his pupils. It is obvious that Chebyshev had an influence on the work of mathematicians who were not his pupils—maybe the clearest example is that of E. I. Zolotarev who almost exclusively worked together with Korkin (see [Ozhi66] and [Ozhi68]), but nevertheless left defining traces in approximation theory.

[2] Comp. section 3.1.

[3] Comp. section 3.4.

[4] Comp. appendix A.1.

[5] Comp. section 3.2.

[6] Comp. section 3.5.

[7] Comp. appendix A.2.

[8] Comp. appendix A.3.

[9] Comp. appendix A.4.

[10] Comp. appendix A.5.

(1855-1932), Andrey Andreevich Markov[11] (1856–1922), Aleksandr Michaylovich Lyapunov[12] (1857–1918), Ivan Ivanovich Ivanov[13] (1862–1939), Dmitri Aleksandrovich Grave[14] (1863–1939), Georgi Feodosevich Voronoy[15] (1868–1908) and Vladimir Andreevich Markov[16] (1871–1897).

Chebyshev and his pupils even had an influence beyond pure mathematics. Maybe the most famous examples are Aleksey Nikolaevich Krylov[17] (1863–1945), shipbuilding engineer and member of the academy, and Dmitri Konstantinovich Bobylev[18] (1842-1917), professor of physics and corresponding member of the academy of sciences. With Bobylev the interaction between mathematics and physics as an example of the realization of the use of mathematics for practical purposes is very clear: one of the opponents of his doctoral thesis was Zolotarev; Bobylev himself was supervisor of Lyapunov's candidate-thesis.

We could of course add others to that list, but we want to restrict ourselves to those scientists who would later be engaged in mathematical research.[19]

---

[11] Comp. section 3.3.1.

[12] Comp. appendix A.6.

[13] Comp. appendix A.7.

[14] Comp. appendix A.8.

[15] Comp. appendix A.9.

[16] Comp. section 3.3.3.

[17] Aleksey Nikolaevich Krylov (*1863, †1945), 1878–1884 Studies at the naval school in St Petersburg, 1884–1888 research at the hydrographical institute, 1888-1890 studies at the naval college (here he visited lectures of A. N. Korkin), since 1890 lecturer there. He also lectured at the polytechnic institute and other colleges, mainly about the theory of shipbuilding. Afterwards Krylov worked (often as their speaker) in several commissions dealing with the development of shipbuilding in Russia and the Soviet Union. He also supervised the construction of ships which were ordered abroad. In 1916 he was elected ordinary member of the academy of sciences, in 1943 he was awarded one of the highest decorations in the Soviet Union and became 'hero of socialistic work.'

[18] Dmitrij Konstantinovich Bobylev (*1842, †1917), in 1862 he finished the college of ordinance, followed by two years of military service, 1864–1870 auditor at St Petersburg university, 1870 candidate thesis in physics, since 1871 lecturer of physics at St Petersburg University and the institute of transportation, 1873 master thesis, 1876 deputy professor (for O. I. Somov) at the chair of mechanics, 1878 extraordinary professor at St Petersburg University for physics, 1885 ordinary professor there.

[19] Chebyshev's pupils N. A. Artemev (1855-1904), Latyshev and Vladimir Vladimirovich Lermantov (1845-?) mentioned in [She94] and [BuJo99] became well-known more as chroniclers than as researchers, e. g., Artemev was the author of the above-cited notes on Chebyshev's lectures on probability theory (see [Erm87]). Lermantov became a physicist—in [Bio96] he was mentioned as an employee of the laboratory of physics at the St Petersburg University (at least from 1870 until 1896). The mathematicians Nikolay Sergeevich Budaev, Mikhail Fedorovich Okatov, Orest Danilovich Khvolson, Iosif Andreevich Kleyber, Yevgeni Vasilevich Borisov, Sergey Yevgenievich Savich, Boris Mikhaylovich Koyalovich

Still long after his death Chebyshev should have an indirect influence on the Russian mathematics because of the work of those pupils who were especially engaged in their common scientific program. A complete list would be confusing because of the large number of different influences. We only want to emphasize Antoni-Bonifatsi Pavlovich Psheborski[20] (1871–1941) and Sergey Natanovich Bernstein[21] (1880-1968), since they were active on approximation theory.

In the following we also will present biographical data of those mathematicians whose efforts were outstanding according to the subject of this work.

3.1 Aleksandr Nikolaevich Korkin

After Chebyshev, Aleksandr Nikolaevich Korkin (1837–1908) was the most important initiator of the formation of the St Petersburg Mathematical School. On the one hand this was due to the fact that among all professors of mathematics he lectured the longest time—22 years—together with Chebyshev at the St Petersburg university. So all the above-mentioned persons who studied there visited both Korkin's and Chebyshev's lectures. We find several sources where the students praised Korkin's efforts as a teacher.

The second reason was the fact that Chebyshev's and his methods are similar to each other. Of course the versatility of Chebyshev's ideas outshone Korkin's work. Nevertheless, Korkin became known as but the founder of Russian mathematical physics.

They agreed with each other in the exclusive use of classical algebraic methods to solve concrete problems and both strove not to lose the connection between mathematics and practice.

3.1.1 About Korkin's Biography

Korkin was born February 19th, 1837 in the village of Zhidovinovo, district of Totem in the Vologda province. He was the son of the state peasant Nikolay Ivanovich Korkin.[22] By his initiative in 1845 the young Aleksandr got the possibility to live and to be educated in Vologda in the house of the grammar school teacher Aleksandr Ivanovich Ivanitski. This was remarkable because at that time in Russia peasants lived in serfdom, Korkin's family was committed to voluntary work for the Russian state («податные»)

---

and Ya. Ya. Tsvetkov might be taken into the list, but either they are known only as lecturers at the St Petersburg University during the interesting time from about 1870 until 1905 or their names are seldom mentioned in comments without any notes about an outstanding connection to the questions discussed here.

[20] Comp. section 5.1.

[21] Comp. section 5.2.

[22] These data are taken from the thorough biography written by E. P. Ožigova [Ozhi68], if we do not mention other sources.

After two years Aleksandr was able to attend the second class of the Vologda grammar school (after his father paid a donation of 200 roubles to the school and of five silver roubles to the Vologda administration to free the son from serfdom).

In 1853 Korkin finished grammar school with the gold medal.[23] In 1854 he registered at the physico-mathematical faculty of St Petersburg University, where at that time mathematics lectures were given by O. I. Somov,[24] Bunyakovski and Chebyshev. Korkin attended Chebyshev's lectures about analytic geometry, higher algebra and number theory.

In 1857 for the first time Korkin was paid attention to because of his contribution 'About Largest and Smallest Quantities' («О наибольших и наименьших величинах»), which was awarded the gold medal in the students' competition. His referee was Bunyakovski. Korkin there investigated several properties of local extrema of explicit or implicit differentiable functions of one or more variables, but he also discussed problems from variational calculus. Especially this subject impressed him.

Because of this outstanding work Korkin was freed from writing a candidate thesis. In 1858, after the final examinations and being freed from serfdom, he could start with his first pedagogical job at the first cadet school (until 1861).

In 1860 a positive report of his work resulted in an offer of an appointment as a lecturer for pure mathematics, which he could start after finishing the master's examinations. On December 11, 1860 he defended his master's thesis 'On the Determination of Abitrary Functions Given by Integrals of Partial Differential Equations' («Об определении произвольных функций в интегралах уравнений с частными производными»). His supervisor was Chebyshev.

In 1861 Korkin's post was confirmed and he became scientific assistant (adjunkt). After the students' unrest in the early summer of 1861 the university was closed for the winter (officially even until August 1863) and the young scientists, including Korkin, were sent abroad 'to prepare the appointment of a professor.'

At first Korkin went to Paris. After a period of self-study on elliptic functions he attended lectures of different mathematicians, among whom were

---

[23] Until now the most outstanding pupils are awarded the gold medal after finishing school. Although some of the persons discussed in this book received the gold medal it was and is a rare decoration and is not awarded in every class.

[24] Osip (Iosif) Ivanovich Somov (1815–1876), until 1835 studies at Moscow University, 1847–1862 ordinary professor at St Petersburg University, 1848–1869 also at the institute of transportation and 1849–1862 at the college of mining, 1857–1862 corresponding, from 1862 ordinary member of the academy. His scientific subjects were theoretical mechanics and analysis. With his works on elliptic functions he laid the foundations for the work of some members of the St Petersburg Mathematical School on this subject.

Liouville and Bertrand. Bertrands lectures about partial differential equations were of special interest for Korkin.

After a brief return to Russia in May 1863, Korkin again left home for Berlin, where he heard Kummer's lectures on circular polynomials, Weierstraß' lectures on elliptic functions and Kronecker's lectures about quadratic forms.

Korkin returned to St Petersburg in September 1864 and again took up his job as a lecturer.

At the end of 1867 he defended his doctoral thesis 'About Systems of Partial Differential Equations of First Order and Some Questions from Mechanics' («О совокупных уравнениях с частными производными 1-го порядка и некоторых вопросах механики»), the opponents again were Chebyshev and Somov.

In May 1868 Korkin became extraordinary professor in the chair of pure mathematics, in 1873 he was promoted to ordinary professor, and in 1886 to merited professor. At the St Petersburg University he lectured until his death in 1908. It is an interesting fact that for thirty years he was the only lecturer to students in advanced courses in partial differential equations and variational calculus [VorSPb] (from winter 1875/76 until 1908). So the whole first generation of the St Petersburg Mathematical School learned these important subjects of mathematical physics only from him and most probably were influenced by him in their opinion about mathematics.

Besides his professorship from 1864 until 1900, Korkin lectured about calculus at the naval college (as a successor of Bunyakovski). Among his students was the later academician Aleksey Nikolaevich Krylov.

3.1.2 The Scientific Work of A. N. Korkin

Korkin's works touch three mathematical branches: the integration of partial differential equations, integration of systems of ordinary differential equations and number theory. The latter work, however, consisted only of joint contributions with his pupil E. I. Zolotarev about quadratic forms.[25] There was only one paper which was connected with approximation theory, „Sur un certain minimum" [KoZo73], written together with Zolotarev in 1873. We will discuss it later in connection with the other works of Zolotarev.

The settings of his problems were exclusively of algebraic nature. He wrote in the introduction into his work „Sur les équations différentielles ordinaires du premier ordre" [Kor96]:

"Recently there were attempts to apply the theory of functions of a complex variable coming itself from investigations about algebraic functions and their integrals to differential equations. But aside of the large generality of these theorems it has another essential imperfectness: as we know it is the disadvantage of its methods to calculate

---

[25] A detailed analysis of Korkin's work would not belong to the aims of the present work. Here we want to refer again to [Ozhi68].

unknown functions. But this calculation is the only true solution of an equation and the definite aim of its analysis. To advance with the integration of differential equations the theory of functions does not suffice; therefore we have to add considerations, which are completely strange from it.

Therefore I think that we have no other chance to reach the aim of the calculation of unknown quantities than to follow the way of the old geometers, that means to restrict oneself to studies of special equations, to search for new equations to integrate; all the more so, since very simple special cases, carefully investigated, might lead to very general conclusions."[26]

In this sense E. P. Ozhigova [Ozhi68] judges:

"In his work about differential equations A. N. Korkin remained within the frames of classical research directions—the search for solutions of differential equations in closed form. He did not accept new methods in this subject."[27]

Here we established an obvious common feature between him and Chebyshev. They rejected 'philosophizing' about infinitely small quantities and always aimed to solve the posed problems with a closed formulae or an algorithm, Korkin was not interested in more theoretical questions about the foundations of analysis and the concept of a function as they recently came up in the middle of the 19th century. Only Moscow mathematicians devoted their work to the analytical theory of differential equations.[28]

---

[26] [Kor96, P. 317]: ,,Dans ces derniers temps on a essayé d'appliquer aux équations différentielles la théorie des fonctions d'une variable complexes, résultant elle même de l'étude des fonctions algébriques et leurs intégrales. Mais, avec la grande généralité de ses théorèmes, elle a aussi une imperfection essentielle: à savoir, le défaut des méthodes pour le calcul des fonctions inconnues. Or, ce calcul est la véritable intégration d'une équation, et le but définitif de son analyse. Pour avancer dans l'intégration des équations différentielles la seule théorie des fonctions ne suffira donc pas; à cet effet il faut y associer des considérations, qui lui sont complèment étrangères.

Je pense donc, qu'ayant pour but le calcul des inconnues nous n'avons jusqu'à présent d'autre moyen que de suivre la marche des ancients géomètres, c'est à dire, en nous bornant à l'étude attentive des équations particulières, rechercher des nouvelles équations intégrables; et cela d'autant plus, que des cas particuliers très simples, traités convenablement, peuvent conduire à des conclusions très générales."

[27] [Ozhi68, S. 60]: «А. Н. Коркин в своих исследованиях по теории дифференциальных уравнений оставался в рамках классического направления - отыскания решений дифференциальных уравнений в конечном виде. Новых методов в этой области он так и не принял.»

[28] Ozhigova mentions Anisimov and Nekrasov.

3.1.3 Judgments about Korkin

As already mentioned above, Korkin played an outstanding role in the formation of the St Petersburg Mathematical School. Although he made almost no contributions to approximation theory, his pedagogical job had an indirect influence on that field. Posse [Pos09] and Krylov [Kry50] emphasized his obviously excellent pedagogical capabilities. Also legendary were the so-called 'Korkin saturdays', where St Petersburg mathematicians and interested pupils met at Korkin's home and discussed mathematical problems. The topics of some theses were developed there.[29]

A conspicuous difference between Korkin and other pupils of Chebyshev is the fact that there was nearly no intersection between his research interests and those of Chebyshev. But, as we saw, they had similar opinions about mathematics.

Korkin, however, seemed to be more rigorous in his rejection of certain mathematical trends than Chebyshev. Posse wrote in his necrologue about Korkin [Pos09]:

> "Korkin extremely negatively related to the direction that mathematics in Germany and partially in France followed in the second half of the 19th century under the influence of Weierstraß and Ricmann. And so he was not interested in the papers of mathematicians of these schools. As someone who somehow loved exaggerations, in this or that direction, he called the above-mentioned direction 'decadency'."[30]

In this prejudice E. P. Ozhigova sees the cause why Sergey Natanovich Bernstein's work had not been acknowledged in St Petersburg. She writes:

> "The academician V. I. Smirnov remembers that after an outstanding defence of his doctoral thesis[31] in Paris S. N. Bernstein returned to St Petersburg to pass the master examinations. Korkin asked him about the integration of partial differential equations with methods of Jacobi and Poisson and was not content with Bernstein's answer. Bernstein had trouble to pass the examinations. He defended his master thesis not in St Petersburg, but in Kharkiv."[32]

---

[29] Zolotarev remembered this.

[30] [Pos09, p. 21]: «К направлению, принятому математикою во вторую половину XIX столетия в Германии и отчасти во Франции, под влиянием Вейерштрасса и Риманна, Коркин относился весьма отрицательно и работами математиков этой школы он не интересовался. Склонный несколько к преувеличениям, в ту и другую сторону, при оценке ученых работ, он назвал вышеупомянутое направление «декаденством».»

[31] The French title 'docteur' was translated here by the Russian word 'doktor' («доктор»). This title, however, rather corresponds to the Russian 'candidate of sciences' («кандидат наук») or the former degree of a 'master' («магистер»). Compare also appendix B.1.

[32] [Ozhi68, p. 53]: «Академик В. И. Смирнов вспоминает, что когда С. Н. Бернштейн, блестяще защитивший в Париже докторскую диссертацию,

3.2 Egor Ivanovich Zolotarev

Referring to what we stated before, Egor Ivanovich Zolotarev is a typical representative of the St Petersburg Mathematical School, that is, someone who aimed to solve his problems completely to the "receipt of a suitable formula or a good algorithm being appropriate for practical computations."[33] Regarding approximation theory it is interesting that he developed a new method to solve extremal problems: he applied methods from the theory of elliptical functions.

3.2.1 Biographical Data

Egor Ivanovich Zolotarev was born March 31, 1847 as a son of Agafya Izotovna Zolotareva and the merchant Ivan Vasilevich Zolotarev in St Petersburg.[34] In 1857 he began to study at the fifth St Petersburg grammar school, a school which centered on mathematics and natural science. He finished it with the silver medal in 1863.[35] In the same year he was allowed to be an auditor («вольнослушатель») at the physico-mathematical faculty of St Petersburg University. He had not been able to become a student before 1864 because he was too young. Among his academic teachers were Somov, Chebyshev and Korkin, with whom he would have a tight scientific friendship.

In November 1867 he defended his candidate thesis 'About the Integration of Gyroscope Equations' and after 10 months there followed his thesis pro venia legendi 'About one question on Minima' [Zol68], which we will discuss later. With this work he was given the right to teach as a 'Privat-dotsent'[36] at St Petersburg University.

First he lectured on 'differential calculus' for students of natural sciences (until summer 1871), later 'integral calculus' and 'introduction to analysis'for beginners of mathematics. Except for a short pause he lectured on 'theory of elliptic functions' for students of advanced courses during his whole job as lecturer and professor [VorSPb].

In December 1869 Zolotarev defended his master thesis 'About the Solution of the Indefinite Equation of Third Degree $x^3 + Ay^3 + A^2z^3 - 3Axyz = 1$.'

приехал в 1906 г. в Петербург держать магистерские экзамены, Коркин задал ему вопрос об интегрировании уравнений в частных производных методами Пуассона и Якоби и остался недоволен его ответом.

Бернштейну с трудом удалось выдержать этот экзамен. Магистерскую диссертацию он защищал не в Петербурге, а в Харкиве.»

[33] [Ozhi66, p. 61]: «получение удобной формулы или хорошего алгоритма, удобного для практики способа вычисления.»

[34] These data were taken from Zolotarev's biography [Ozhi66].

[35] According to S. Ya. Grodzenski [Gro87], the biographer of Andrey Andreevich and Vladimir Andreevich Markov, aside from Zolotarev Chebyshev's successors A. V. Vasilev, A. A. and V. A. Markov were pupils of the fifth grammar school, so we can guess the significance this school had at that time.

[36] A lecturer who is not a member of the salaried university staff.

He started his first trip abroad in 1872, visiting Berlin and Heidelberg. In Berlin he attended Weierstraß' "theory of analytic functions," in Heidelberg Königsberger's "theory of functions of a complex variable."

In 1874 Zolotarev became a member of the university staff as a lecturer and in the same year he defended his doctoral thesis 'Theory of Complex Numbers with an Application to Integral Calculus.' The problem Zolotarev solved there, was based on a problem Chebyshev had posed before, the representation of expressions of the form

$$\int_a^b \frac{x + A}{\sqrt{x^4 + ax^3 + bx^2 + cx + d}}$$

by logarithms. This was a question Chebyshev had been interested in since the beginning of his research,[37] but he was not able to solve it without the help of elliptic functions.[38]

With the beginning of the winter semester in 1876, Zolotarev was appointed extraordinary professor and after the death of academician Somov he became his successor, but only as an adjunct of the academy of sciences. Nevertheless this was a remarkable fact. On the one hand there was another candidate, the ordinary professor Korkin.[39]

Egor Ivanovich Zolotarev's steep career ended abruptly with his early death. On June 26th, 1878, when he was on his way to the dacha he was run over by a train in the station Tsarskoe Selo (now Pushkin). On July 7th, 1878 he died from blood-poisoning.

3.2.2 Application of the Theory of Elliptic Functions to Approximation Theory

In his thesis pro venia legendi 'About one question on Minima' [Zol68] Zolotarev discussed Chebyshev's first problem for two given coefficients, that is for an arbitrary but fixed $\sigma \in \mathbb{R}$ he tried to solve

$$\min_{p \in \mathbb{P}_{n-2}} \max_{x \in [-1,1]} |x^n - \sigma x^{n-1} - p(x)|. \tag{3.1}$$

This problem was posed to him by Chebyshev, as Zolotarev himself would mention[40] later in his 1877's work 'Application of Elliptic Functions ...' [Zol77/1], published also in French [Zol78]).

---

[37] Compare, e. g. the topic of [Cheb47].

[38] Compare the discussion in [Erm94/1].

[39] Korkin should then become Zolotarev's successor after his death, but was not elected because he was professor of 'pure mathematics,' but Zolotarev's position was that of 'applied mathematics.' That Korkin had not been elected, caused some excitement (see [Ozhi68, S. 50ff.]).

[40] "Ten years ago I was recommended by P. L. Chebyshev to investigate this problem" [Zol32, Vol. 2, p. 2]: «Десять лет тому назад этот вопрос был мне рекомендован для занятий П. Л. Чебышевым [...] ».

To solve this problem he at first proved theorems of Chebyshev type (compare Theorem 2.1 from page 38), where the assumptions were even a little more general, since Zolotarev demanded that the parameter of the problem, $p_1, \ldots, p_n$, should satisfy ν different side-conditions

$$\phi_1(p_1, \ldots, p_n) = 0$$

$$\vdots$$

$$\phi_\nu(p_1, \ldots, p_n) = 0.$$

Then, as before with Chebyshev, there follows a theorem about the number of deviation points of the error function

$$F(x) = Y(x) - \sum_{i=0}^{n-1} p_i x^i, \quad x \in [-h, h].$$

Here Zolotarev did not assume anything for the choice of the function Y. It is an 'arbitrary' function.

Theorem 3.1 *Let* $L := \max\limits_{x \in [-h,h]} |F(x)|$. *If* F *satisfies*

$$F(a_1) = A_1$$

$$\vdots$$

$$F(a_m) = A_m,$$

then there holds:

If among all functions satisfying the above-described side-conditions F *is the one which deviates the least possible from zero, then the equations*

$$F(x)^2 - L^2 = 0 \quad and$$
$$(x^2 - h^2)F'(x) = 0$$

have at least $n + 1 - m$ *common roots.*

We remark that the number of deviation points gets smaller with the number of the side-conditions. We could make such an observation already according to the other cases Chebyshev discussed (weighted and rational approximation).

Zolotarev used this theorem to write down the characteristic equations for his original problem with two given coefficients, that is, for

$$Y(x) := x^n$$

and

$$\phi_1(p_1, \ldots, p_n) = p_n + \sigma.$$

To simplify the further calculations he set the borders of the interval equal to -1 and 1.

In the case that the characteristic equations are solved in only one border, then the characteristic equations can be written as

$$F(x)^2 - L^2 = (x \pm 1)(x - \alpha) \prod_{i=1}^{n-1} (x - x_i)^2 \quad \text{and} \tag{3.2}$$

$$F'(x) = n \prod_{i=1}^{n-1} x - x_i.$$

$x_1, \ldots, x_{n-1}$ are then points within $[-1, 1]$, and α might be a point beyond the interval.

In the case that the characteristic equations are solved at both borders, they have the form

$$F(x)^2 - L^2 = (x^2 - 1)(x - \alpha)(x - \beta) \prod_{i=1}^{n-2} (x - x_i)^2 \quad \text{and} \tag{3.3}$$

$$F'(x) = \rho(x) \prod_{i=1}^{n-1} x - x_i,$$

where ρ is a linear function. Comparing the coefficients with $F(x) = x^n - \sigma x^{n-1} + \cdots$ we get

$$\rho(x) = n \left(x - \frac{\alpha + \beta}{2} \right) + \sigma.$$

3.2.2.1 Zolotarev's solutions

With long and difficult calculations based on the theory of elliptic functions he was able to determine the solutions of the problem.

In the first case (3.2), which can occur for $\sigma < n \tan^2 \frac{\pi}{2n}$, the solution is

$$F(x) = \frac{1}{2^n} \left(x - \frac{\sigma}{n} + \sqrt{(x+1)\left(x - 1 - \frac{2\sigma}{n}\right)} \right)^n \tag{3.4}$$

$$+ \left(x - \frac{\sigma}{n} - \sqrt{(x+1)\left(x - 1 - \frac{2\sigma}{n}\right)} \right)^n$$

with the maximum error

$$L = \frac{\left(\frac{\alpha \pm 1}{2}\right)^n}{2^{n-1}}. \tag{3.5}$$

The sign is equal to the sign of the given quantity σ.

The second case (3.3) has to be divided into the cases of the roots α and β being real or imaginary numbers.

If there holds $\sigma > n \tan^2 \frac{\pi}{2n}$, then both of these numbers can be real. Then the solution is[41]

$$\frac{F(x)}{L} = \frac{1}{2}\left[\left(\frac{H\left(\frac{K}{n}+u\right)}{H\left(\frac{K}{n}-u\right)}\right)^n + \left(\frac{H\left(\frac{K}{n}-u\right)}{H\left(\frac{K}{n}+u\right)}\right)^n\right]; \qquad (3.8)$$

the parameter k of the elliptic function can be determined by

$$1 + \frac{\sigma}{n} = \frac{2\,\mathrm{sn}\,\frac{K}{n}}{\mathrm{cn}\,\frac{K}{n}\,\mathrm{dn}\,\frac{K}{n}}\left[\frac{1}{\mathrm{sn}\,\frac{2K}{n}} - \frac{\Theta'\left(\frac{K}{n}\right)}{\Theta\left(\frac{K}{n}\right)}\right].$$

The maximum error is

$$L = \frac{(-1)^n}{2^{n-1}}\left[\frac{k\Theta_1^2(0)}{H_1\left(\frac{K}{n}\right)\Theta_1\left(\frac{K}{n}\right)}\right]^n. \qquad (3.9)$$

If σ is even larger than n, then this solution is unique.

If both roots α and β are purely imaginary, then we get the solution

$$\frac{F(x)}{L} = \frac{1}{2}\left[\left(\frac{H\left(\frac{K}{n}+u\right)\Theta_1\left(\frac{K}{n}+u\right)}{H\left(\frac{K}{n}-u\right)\Theta_1\left(\frac{K}{n}-u\right)}\right)^n + \left(\frac{H\left(\frac{K}{n}-u\right)\Theta_1\left(\frac{K}{n}-u\right)}{H\left(\frac{K}{n}+u\right)\Theta_1\left(\frac{K}{n}+u\right)}\right)^n\right]; \qquad (3.10)$$

---

[41] We want to remember the definition of the elliptic functions which occur in the following formulae.

The amplitude ϕ is defined by the integral

$$x = \int_0^\phi \frac{d\alpha}{\sqrt{1 - k^2 \sin^2 \alpha}}$$

as is attributed by $\phi = \mathrm{am}\,x$. Then the elliptic functions sn, cn and dn are defined by

$$\mathrm{sn}\,x := \sin\mathrm{am}\,x := \sin\phi,$$

$$\mathrm{cn}\,x := \cos\mathrm{am}\,x := \cos\phi \quad \text{and}$$

$$\mathrm{dn}\,x := \Delta\,\mathrm{am}\,x := \sqrt{1 - k^2 \sin^2 \phi}.$$

Today the Jacobian functions H, H_1, Θ and Θ_1 are denoted as $\vartheta_1, \ldots, \vartheta_4$. Here it is

$$H(x) = \vartheta_1\left(\frac{x}{2K}\right), \quad \Theta(x) = \vartheta_2\left(\frac{x}{2K}\right), \qquad (3.6)$$

$$H_1(x) = \vartheta_3\left(\frac{x}{2K}\right), \quad \Theta_1(x) = \vartheta_4\left(\frac{x}{2K}\right). \qquad (3.7)$$

the parameter k of the elliptic function can be determined by

$$\frac{\sigma}{n} = \frac{\operatorname{sn}\frac{2K}{n}}{\operatorname{dn}\frac{2K}{n}}\left[\frac{k^2\operatorname{sn}\frac{2K}{n}\operatorname{cn}^2\frac{K}{n}}{\operatorname{dn}\frac{2K}{n}} - 2\frac{\Theta'\left(\frac{K}{n}\right)}{\Theta\left(\frac{K}{n}\right)}\right].$$

The maximum error is

$$L = \frac{(-1)^n}{2^{n-1}}\left[\frac{\Theta_1(0)}{\Theta_1\left(\frac{2K}{n}\right)}\right]^n. \tag{3.11}$$

This case can occur only if σ satisfies the inequality

$$\frac{\operatorname{cn}\frac{2K}{n}}{\operatorname{dn}^2\frac{2K}{n}} < 1 + \frac{\sigma}{n}.$$

Looking at the solutions we can guess how complicated Zolotarev's calculations were. All the methods of Jacobi and Abel according to elliptic functions were used by him.

Figure 3.1 shows in the simple case (without elliptic functions), how the error changes by choosing the second coefficient σ.

Zolotarev could show that

$$\lim_{\sigma \to -\infty} L(\sigma) = \infty.$$

Zolotarev deepened this method in his second work about approximation theory, 'Application of Elliptic Functions ...' [Zol77/1]. There he simplified the calculations from [Zol68] and solved another interesting extremal problem, the determination of the fraction

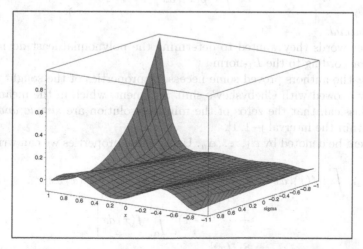

Figure 3.1. *The figure shows Zolotarev's solution in dependence on σ. It can be seen clearly that the maximum error enormously grows when σ becomes negative.*

$$y(x) := \frac{\phi(x)}{\psi(x)}, \quad \phi \in \mathbb{P}_n, \psi \in \mathbb{P}_n$$

of maximal deviation from zero on the half axis $(\frac{1}{k}, \infty)$, $k < 1$, if additionally:

$$|y(x)| \leq 1, \quad |x| < 1.$$

Similar to this problem is the determination of the fraction

$$y(x) := \frac{\phi(x)}{\psi(x)}, \quad \phi \in \mathbb{P}_n, \psi \in \mathbb{P}_n$$

of least deviation from zero in the interval $[-1, 1]$, if there holds beyond this interval:

$$y(x) \leq -1, \quad -\frac{1}{k} \leq x \leq -1,$$

$$y(x) \geq 1, \quad 1 \leq x \leq \frac{1}{k}.$$

3.2.3 L_1-Approximation

The following problem was solved by Korkin and Zolotarev in their joint work Sur un certain minimum" [KoZo73]:

To determine the coefficients of the polynomial

$$p(x) := x^n + a_{n-1}x^{n-1} + \cdots + a_0,$$

so that

$$\int_{-1}^{1} |p(x)| \, dx$$

will be minimal.

In other words they wanted to determine the polynomial least deviating from zero according to the L_1-norm.

At first the authors proved some necessary properties of the sought solution. They showed with Chebyshev's shift argument, which in the meantime became classical, that the zeros of the minimal solution are *simple and real* and lie within the interval $[-1, 1]$.

Let them be denoted by $\alpha_1, \ldots, \alpha_n$. Using these properties we can write:[42]

$$\int_{-1}^{1} |f(x)| \, dx = (-1)^n \left[\int_{-1}^{\alpha_1} f(x) \, dx - \int_{\alpha_1}^{\alpha_2} f(x) \, dx \right. \tag{3.12}$$

$$\left. + \cdots + (-1)^n \int_{\alpha_n}^{1} f(x) \, dx \right]$$

$$=: S[f(x)].$$

---

[42] Later they would show that indeed the sign is $(-1)^n$.

The function $S : \mathcal{A} \to \mathbb{R}$ as defined above can be regarded as a function of the unknown coefficients $S(f) = S(a_0, \ldots, a_{n-1})$, where ($\mathcal{A} \subset \mathbb{P}_n[-1, 1]$ denotes the polynomials with simple real zeros which lie within the interval $[-1, 1]$.

A minimum of this function will satisfy the necessary condition

$$\frac{\partial S}{\partial a_0}(a_0, \ldots, a_{n-1}) = \cdots = \frac{\partial S}{\partial a_0}(a_0, \ldots, a_{n-1}) = 0,$$

so for $i = 0, \ldots, n - 1$:

$$
\begin{aligned}
0 &= \frac{\partial S}{\partial a_i}(a_0, \ldots, a_{n-1}) \\
&= \frac{(-1)^n}{i+1}\left[\int_{-1}^{\alpha_1} x^{i+1}\,dx - \int_{\alpha_1}^{\alpha_2} x^{i+1}\,dx + \cdots + (-1)^n \int_{\alpha_n}^{1} x^{i+1}\,dx\right] \\
&= \frac{(-1)^n}{i+1}\left[2\alpha_1^{i+1} - 2\alpha_2^{i+1} + \cdots + 2(-1)^n\alpha_n^{i+1} + [(-1)^n + (-1)^{i+2}]\right].
\end{aligned}
$$

Thus, the necessary condition for the zeros of the minimal solution will be

$$\alpha_1^{i+1} - \alpha_2^{i+1} + \cdots + (-1)^n \alpha_n^{i+1} = \frac{(-1)^{i+1} + (-1)^{n+1}}{2}. \tag{3.13}$$

Before calculating the minimal solution two further properties are shown:

At first the numbers $\alpha_1, \ldots, \alpha_n$ determined by (3.13) and the previous properties are *uniquely* determined. So the necessary condition is sufficient as well.

Secondly the following functional equation holds for the minimal solution f:

$$f(x) = (-1)^n f(-x). \tag{3.14}$$

So, f and n are both at the same time either even or odd. Therefore the sign in (3.12) is correct.

With the equations (3.13) the solutions are determined for $n = 1, 2, 3, 4$. They are

$$f_1(x) = x$$
$$f_2(x) = x^2 - \frac{1}{4} \quad [= (x + \frac{1}{2})(x - \frac{1}{2})]$$
$$f_3(x) = x^3 - \frac{1}{2}x \quad [= (x + \frac{1}{\sqrt{2}})x(x - \frac{1}{\sqrt{2}})]$$
$$f_4(x) = x^4 - \frac{3}{4}x^2 + \frac{1}{16}$$
$$[= (x + \frac{1 + \sqrt{5}}{4})(x - \frac{1 - \sqrt{5}}{4})(x + \frac{1 - \sqrt{5}}{4})(x - \frac{1 + \sqrt{5}}{4})].$$

This heuristic approach led Korkin and Zolotarev to the proposition for the general solution:

$$f_n(x) = \frac{1}{2^n} \frac{\sin((n+1)\arccos x)}{\sqrt{1-x^2}}.\tag{3.15}$$

In the following sections of Korkin's and Zolotarev's paper, it is going to be verified.

Today we know the functions f_n as *Chebyshev polynomials of second kind,* they are connected with the Chebyshev polynomials of first kind by the property

$$f_n(x) = \frac{1}{n+1}T'_{n+1}(x),$$

which also shows that all the f_n are polynomials.[43]

3.2.3.1 An excursion into Laurent coefficients

The aim of this excursion is to have the necessary conditions generally handier. So a connection between the zeros of a polynomial and the Laurent expansion of certain rational functions will be established.

Let u be a function given by the main part of a Laurent expansion

$$u(x) := \sum_{i=1}^{\infty} a_{-i}x^{-i}, \quad a_{-1} \neq 0.\tag{3.16}$$

The polynomial $\phi \in \mathbb{P}_{n+1}$ has to be chosen so that the coefficients $a'_{-1}, \ldots, a'_{-(n+1)}$ of the expansion of $u\phi$ vanish, hence:

$$u(x)\phi(x) := \sum_{i=n+2}^{\infty} a'_{-i}x^{-i}.\tag{3.17}$$

Then we can find functions ψ and ε with

$$u(x)\phi(x) = \psi(x)(1 + \varepsilon(x))\tag{3.18}$$

where $\psi \in \mathbb{P}_n$ and

$$\varepsilon(x) = \sum_{i=2n+2}^{\infty} b_{-i}x^{-i}.\tag{3.19}$$

The function ψ will later be the minimal solution. We assume that its zeros are all different from each other and real. We denote them by

$$c_1, \ldots, c_n.$$

Then we can expand:

---

[43] The polynomials nowadays used differ from the original ones by the factor $\frac{1}{2^{n-1}}$, which was used earlier to get 1 as the first coefficient.

$$\frac{\phi(x)}{x - c_i} = \hat{\phi}(x) + \sum_{j=1}^{\infty} \frac{c_i^{j-1}\phi(c_i)}{x^j}$$

with $\hat{\phi} \in \mathbb{P}_n$ or completely:

$$\sum_{i=1}^{n} \frac{\phi(x)}{x - c_i} = \phi(x)\frac{\psi'(x)}{\psi(x)} \tag{3.20}$$

$$= \tilde{\phi} \sum_{j=1}^{\infty} \frac{\sum_{i=1}^{n} c_i^{j-1}\phi(c_i)}{x^j}, \quad \tilde{\phi} \in \mathbb{P}_n.$$

Deriving both sides of (3.18) we get

$$u'(x)\phi(x) + u(x)\phi'(x) = \psi'(x)(1 + \varepsilon(x)) + \psi(x)\varepsilon'(x),$$

by dividing by (3.18) it becomes

$$\frac{u'(x)}{u(x)} + \frac{\phi'(x)}{\phi(x)} = \frac{\psi'(x)}{\psi(x)} + \frac{\varepsilon'(x)}{1 + \varepsilon(x)}.$$

This equation can also be written as

$$\phi'(x) + \phi(x)\frac{u'(x)}{u(x)} = \phi(x)\frac{\psi'(x)}{\psi(x)} + \phi(x)\frac{\varepsilon'(x)}{1 + \varepsilon(x)}.$$

Since all coefficients $\geq -(n+1)$ vanish in the Laurent expansion of $\phi\frac{\varepsilon'}{1+\varepsilon}$, the coefficients for $x^{-1}, \ldots, x^{-(n+1)}$ of the expansion of

$$\phi(x)\frac{u'(x)}{u(x)} \quad \text{and} \quad \phi(x)\frac{\psi'(x)}{\psi(x)}$$

must coincide.

But with (3.20) they are

$$a_{-1} := \sum_{i=1}^{n} \phi(c_i), \ldots, a_{-(n+1)} := \sum_{i=1}^{n} c_i^n \phi(c_i).$$

So it is now possible to choose suitable functions. Set

$$u(x) := \frac{1}{\sqrt{x^2 - 1}},$$

$$\phi(x) := \frac{1}{2^n}\cos[(n+1)\arccos(x)] \quad \left(= \frac{1}{2^n}T_{n+1}(x)\right) \quad \text{and}$$

$$\psi(x) := \frac{1}{2^n}\frac{\sin[(n+1)\arccos(x)]}{\sqrt{1 - x^2}} \quad (= f_n(x)).$$

As we have already shown, ϕ and ψ are polynomials. Obviously their zeros satisfy the demanded conditions.

Then

$$\phi(x)\frac{u'(x)}{u(x)} = -\frac{x}{x^2-1}\phi(x).$$

Because (x^2-1) is not a factor of ϕ, ϕ can be represented as

$$\phi(x) = (x^2-1)F(x) + Ax + B \tag{3.21}$$

with $F \in \mathbb{P}_{n-1}$ and non-vanishing constants A and B. Putting in 1 and -1 we get

$$A = \frac{1-(-1)^{n+1}}{2^{n+1}} \quad B = \frac{1+(-1)^{n+1}}{2^{n+1}}. \tag{3.22}$$

With (3.21) we have then

$$\phi(x)\frac{u'(x)}{u(x)} = -xF(x) - A - \frac{Bx+A}{x^2-1} \tag{3.23}$$

$$= -\left(A + xF(x) + \frac{B}{x} + \frac{A}{x^2} + \frac{B}{x^3} + \cdots\right).$$

Now we set

$$\alpha_i := \cos\left(\frac{(n-i+1)\pi}{n+1}\right), \quad i = 1, \ldots, n,$$

as zeros of ψ, for which property (3.13) has to be verified. So we reached by definition:

$$\alpha_1 < \alpha_2 < \cdots < \alpha_n,$$

and compute

$$\phi(\alpha_i) = \frac{1}{2^n}(-1)^{n+i+1}.$$

The result of the previous section, the general representation of the Laurent coefficients (3.20) and the concrete calculation (3.23) now produce for all $i = 1, \ldots, n$, that the sum

$$\sum_{i=1}^{n} \alpha_i^j \phi(\alpha_i) = -\frac{1}{2^n}(-1)^{n+1}\sum_{i=1}^{n}(-1)^{i-1}\alpha_i^j$$

for even j will be equal to B and for odd j will be equal to A.[44]

So it must be:

$$-\frac{1}{2^n}(-1)^{n+1}\sum_{i=1}^{n}(-1)^{i-1}\alpha_i^j = -\frac{1+(-1)^{n+1+j}}{2^{n+1}},$$

---

[44] The exponent j corresponds with the Laurent coefficient $j+1$.

for the right expression for even j will assume B and for odd j assume A.
Multiplying both sides with $-\frac{1}{2^n}(-1)^{n+1}$ will lead to

$$\frac{\partial S}{\partial a_{j-1}} \sum_{i=1}^{n}(-1)^{i-1}\alpha_i^j = \frac{(-1)^j + (-1)^{n+1}}{2},$$

and is equal to condition (3.13), which had to be verified.

So ψ is the minimal solution to be determined.

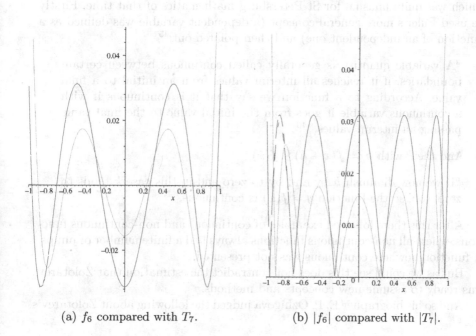

(a) f_6 compared with T_7.   (b) $|f_6|$ compared with $|T_7|$.

Figure 3.2. *The figure shows Korkin's and Zolotarev's solution for n=6, which is compared with T_7. Especially the right picture presenting their absolute values clearly shows that $\int_{-1}^{1}|f_6(x)|\,dx < \int_{-1}^{1}|T_7(x)|\,dx$.*

The application of methods from the theory of functions is remarkable and could not be found in the work of Chebyshev.

Here the influence of O. I. Somov, who represented the theory of elliptic functions among the teachers of St Petersburg University, became clear. In 1876 Zolotarev wrote his necrologue [Zol77/2], where he expressed this.

So Zolotarev proved to be ready to use new (Western European) results to answer questions of determining maxima and minima of concrete practical problems dominating in St Petersburg.

3.2.4 Zolotarev's Conceptional Apparatus

In his contributions to approximation theory Zolotarev usually did not exactly define what functions he was talking about, although this would have been an interesting question, e. g., for Theorem 3.1.

His lectures, however, showed that he was interested in questions about the foundations of analysis.

So already in his first lectures "Differential Calculus for Students of Natural Science" (since 1868) he discussed the conception of a function in a way which was quite unusual for St Petersburg mathematics of that time. Firstly he used Euler's more general concept (a dependent variable was defined as a function of an independent one) and then pointed out:[45]

> "A variable quantity is generally called continuous between certain boundaries, if it reaches all interim values from an initial to a final value. According to a function we say that it is continuous if with a continuous variable it goes from the initial value to the final value passing all interim values."[46]

And then, with $k := f(x + h) - f(x)$:

> "If we now diminish h by and by to zero and in this case k tends to zero, too, so the function $y = f(x)$ is continuous."[47]

After this there followed examples of continuous and non-continuous functions, where all non-continuous functions always had a finite number of jumps. A function nowhere continuous was not presented.

But as we could see this does not contradict the estimation that Zolotarev was ready to assume new concepts and methods.

And so his biographer E. P. Ozhigova judged the following about Zolotarev's work:

> Although Zolotarev was a pupil of Chebyshev, his work had a set of characteristic particularities differing from the usual methods of the mathematics of the St Petersburg school. [...] He often used the theory of elliptic functions which Chebyshev did not love much. Whereas for Chebyshev the solution of concrete problems and the formulation

---

[45] The cited transcription was made in 1870.

[46] [Zol70, p. 10]: «Переменная величина называется вообще непрерывною между некоторыми пределами, если она переходит от начального значения к конечному через все промежуточные значения. Относительно функтсии говорят, что она непрерывна, если с непрерывным изменением независимой переходит от начального значения к конечному через все промежуточные величины.»

[47] [Zol70, p. 10]: «Если теперь станем h последовательно уменьшать до нуля и в этом случае k будет тоже стремиться к нулю, то функция $y = f(x)$ непрерывна.»

of new problems based on certain practical results were charateristic, for Zolotarev it was the pursuit to abstraction and generalization."[48]

This became especially clear when analysing Zolotarev's notebooks from his assets where he formulated his future scientific aims. Often we find an additional aim to the idea of solving a concrete problem: to generalize the discovered theorems.[49] Chebyshev never posed such problems!

Regarding approximation theory, we can confirm that Zolotarev was the first who applied function theory there, a non-typical subject for St Petersburg mathematicians.

That Zolotarev did not discuss the more general problem of approximation of continuous or integrable functions or did not look for generalizations of Chebyshev's theorems, is not really a flaw because modern analysis had not prepared all the basic methods to do this.

Zolotarev had an excellent command of the well developed methods from algebra and function theory, as his main monographs 'Theory of integral complex numbers' (today we say 'integral algebraic numbers') and the abundance of contributions to the theory of quadratic forms (written together with Korkin) show.

Because of his very early death Zolotarev did not experience the further development of the foundations of mathematical analysis and so could not take part in it. His lectures where he discussed the concept of continuity proved that he obviously would have been interested in it.

3.3 Andrey and Vladimir Andreevich Markov

Among all of Chebyshev's pupils, Andrey Andreevich Markov is regarded as the one who is the most similar to Chebyshev with respect to the amount of mathematical work, aims and methods.

N. I. Akhiezer summarized Markov's work as reprinted in the "Selected Works about the Theory of Continuous Fractions and the Theory of Functions Least Deviating from Zero" [MarA48]:

[...] There starts a whole set of papers devoted to limits of integrals, interpolation, functions of least deviation from zero and continuous

---

[48] [Ozhi66, p. 121:]«Хотя Золотарев был учеником Чебышева, его творчество имеет ряд характерных особенностей, отличающих его от обычных методов математиков Петербургской школы. [...] Он часто пользуется теорией эллиптических функций, которую не очень любил Чебышев. Если для Чебышева характерно решение конкретных задач на основе некоторых практических соображений [...], то для Золотарева характерно стремление к абстракции, к обобщениям.»

[49] See [Ozhi66, Chapter 9].

fractions—works, where Markov turned out to be the successor of his ingenious teacher P. L. Chebyshev."[50]

The same judgement holds for probability theory, since especially the Markov strings made him famous.

3.3.1 About Andrey Andreevich Markovs Life

Andrey Andreevich Markov was born June 2, 1856 in Ryazan as the son of the secretretary of the public forest management of Ryazan, Andrey Grigorevich Markov, and his first wife, Nadezhda Petrovna Markova.[51]

In the beginning of the 1860s Andrey Grigorevich moved to St Petersburg to adopt the job as an asset manager of the princess Ekaterina Aleksandrovna Valvateva.

In 1866 Andrey Andreevich's school life began with his entrance into St Petersburg's fifth grammar school. Already during his school time Andrey was intensely engaged in higher mathematics. As a 17 years old grammar school pupil he informed Bunyakovski, Korkin and Zolotarev about an apparently new method to solve linear ordinary differential equations and was invited to the Korkin Saturdays. At last in 1874 he finished the school and began his studies at the physico-mathematical faculty of St Petersburg University.

Among his teachers were[52] Sochocki (differential calculus, higher algebra), Posse (analytic geometry), Zolotarev (integral calculus), Chebyshev (number theory, probability theory), Korkin (ordinary and partial differential equations), Okatov (mechanism theory), Somov (mechanics) and Budaev (descriptive and higher geometry).

In 1877 he was awarded the gold medal for his outstanding[53] solution of the problem "About Integration of Differential Equations by Continuous Fractions with an Application to the Equation $(1 + x^2)\frac{dy}{dx} = n(1 + y^2)$." In the following year he passed the candidate examinations and remained at the university to prepare for the lecturer's job.

In April 1880, Andrey Markov defended his master thesis "About Binary Quadratic Forms with Positive Determinant," which was encouraged by Korkin and Zolotarev. It was reported by Chebyshev, Korkin and Sochocki.

Another five years later, in January 1885, there followed his doctoral thesis "About Some Applications of Algebraic Continuous Fractions," where he discussed the following moment problem: To find a function f satisfying

---

[50] [Akh48, S. 10]: «[...] начинается целый цикл работ Маркова, посвященных предельным величинам интегралов, интерполированию, функциям, наименьшего уклонения от нуля и непрерывным дробям - работ, в которых Марков явился продолжателем своего гениального учителя П. Л. Чебышева.»

[51] The biographical data was taken from [Gro87].

[52] Comp. [VorSPb].

[53] Due to Zolotarev's judgment.

$$\int_a^b x^k f(x)\,dx = c_k, \quad k = 0, \dots, n.$$

For a given function g Markov solved there the problem to determine an upper bound for the functional

$$J(f) := \int_a^b f(x)g(x)\,dx.$$

His pedagogical job began after the defence of his master thesis in autumn 1880. As a privat-dotsent he lectured on differential and integral calculus. Later he lectured alternately "introduction to analysis," probability theory (succeeding Chebyshev who had left the university in 1882) and calculus of differences. From 1895/96 until 1905 he additionally lectured again differential calculus.[54]

One year after the defence of the doctoral thesis he was appointed extraordinary professor (1886) and in the same year he was elected adjunct of the academy of sciences. In 1890, after the death of the academician Bunyakovski, Markov became extraordinary member of the academy. His promotion to an ordinary professor of St Petersburg University followed in autumn 1894.

Finally in 1896 he was elected ordinary member of the academy as the successor of Chebyshev. In 1905 he was appointed merited professor and got the right to retire which he immediately used. Until 1910, however, he continued to lecture on calculus of differences.

Protesting against a decree of the ministry of education Markov rejected a further teaching activity at the St Petersburg University. In connection with students' riots in 1908, professors and lecturers of St Petersburg University were ordered to maintain observations over their students. Firstly Markov rejected this decree and after that he wrote an explanation where he expressed his refusal to be an "agent of the governance." Finally he saw no recourse other than full retirement from the university.

In 1913 the council of St Petersburg elected nine scientists honorary members of the university. Markov was among them, but his election alone was not affirmed by the minister of education. The affirmation was done only four years later, after the February revolution in 1917.

In the following year Markov again resumed his teaching activity. Until his death in 1922 he lectured on probability theory and calculus of differences.

3.3.2 The Early Work on Approximation Theory

In his work 'Determination of a Function with Respect to the Condition to Deviate as Little as Possible from Zero' [MarA84], Markov discussed a generalization of the second case from ,,Sur les questions de minima [...]" [Cheb59], the weighted approximation problem. To be precise, he presented the problem

---

[54] See [VorSPb].

$$\min_{p \in \mathbb{P}_{n-1}} \max_{x \in [-1,1]} \left| \frac{x^n - p(x)}{\sqrt{q(x)}} \right|, \quad q \in \mathbb{P}_{2n}. \tag{3.24}$$

With respect to methods developed by Zolotarev in [Zol77/1] and his own trigonometric approach he was able to solve the problem without the use of the complex continuous fractions Chebyshev needed. Markov introduced the differential equation[55]

$$d(\iota^n - p) = 0.$$

Such an approach had already been succesful for Chebyshev's first case (approximation by polynomials) in ,,Théorie des mécanismes [...]" [Cheb54]. Here it could be generalized.

3.3.2.1 About a Question by D. I. Mendeleev. An Alternation Theorem

In the paper [MarA90] A. A. Markov formulated the following problem which had been posed before by Dmitri Ivanovich Mendeleev for the case $n = 2$ in his work ,,Investigation of Aqueous Dilutions according to the Specific Weight."[56] He was looking for the smallest upper bound for the derivative of a polynomial with given norm. With this we define[57]

$$\mathbb{P}_{n,L} = \{p \in \mathbb{P}_n \mid \|p\| = L\}. \tag{3.25}$$

Using this definition we want to get an approach for two problems. Firstly, for given $x \in \mathbb{R}$ we have to determine the quantity

$$\max_{p \in \mathbb{P}_{n,L}} |p'(x)| \tag{3.26}$$

and the second problem is the determination of an upper bound for the norm of the first derivative of a polynomial depending on the norm of the polynomial itself.

3.3.2.1.1 First Problem

A. Markov's approach began with the following definitions:

- Let $z \in \mathbb{R}$ be given. $\hspace{6em}$ (3.27)
- Let $p_0 \in \mathbb{P}_{n,L}$ be chosen as follows: $|p_0'(z)| \geq |q'(z)| \, \forall q \in \mathbb{P}_{n,L}$. $\hspace{1em}$ (3.28)

Then he paid attention to the deviation points of p_0 and formulated the following theorem [MarA90, p. 52]:

---

[55] We want to use modern notation.

[56] A. Markov wrote in [MarA90, S. 51]: «Такой вопрос поставлен Д. И. Менделеевым при $n = 2$ в его сочинении «Исследование водных растворов по удельному весу (§86)».

[57] The formulation is not due to A. Markov, we use it to abbreviate.

Theorem 3.2 (Necessary alternation criterion by A. A. Markov)
Let $a \le \alpha_1 < \ldots < \alpha_s \le b$ be the deviation points of the polynomial p_0 defined in (3.28), that is,

$$|p_0(\alpha_1)| = |p_0(\alpha_2)| = \cdots = |p_0(\alpha_s)|.$$

Then there holds: At least $n - 1$ of the quotients

$$\frac{p_0(\alpha_2)}{p_0(\alpha_1)}, \quad \frac{p_0(\alpha_3)}{p_0(\alpha_2)}, \quad \ldots, \quad \frac{p_0(\alpha_s)}{p_0(\alpha_{s-1})} \tag{3.29}$$

are equal to -1.

More clearly, we can say: *Among the quotients (3.29) there is a Chebyshev $(n-1)$-alternant.*

But because of the special character of Markov's problem defined by the side-condition (3.28) he did not formulate an exact alternation theorem. But with this theorem the existence of an $(n - 1)$-alternant was proven for a minimal solution, too. So, only two alternations missed. Markov's proof was constructive and used the same arguments which are still used to prove the alternation theorem elementarily (e. g., in [Nat49]).

Proof Assume that p_0 had less alternations, let's say only $k \le n - 2$.

Then we find a polynomial $q \in \mathbb{P}_{n-2}$, whose zeros lie only between these alternations and for which holds:

$$\text{sign } q(\alpha_i) = -\text{sign } p_0(\alpha_i) \, \forall i = 1, \ldots, s. \tag{3.30}$$

Now we construct a polynomial P with

$$P(x) = p_0(x) + \varepsilon(x - z)^2 q(x).$$

Then:

- $P \in P_n$
- $\|P\| < \|p_0\| = L$ for a suitable[58] $\varepsilon > 0$
- $p'(z) = p_0'(z)$

And so there holds for $Q := \frac{L}{\|P\|}$:

---

[58] For a suitable choice of ε we imagine a subdivision of $[a, b]$ into segments, where the variation of p_0 does not exceed L. For extremal segments containing an extremal point then there holds

$$\varepsilon < \frac{L}{2\|\iota - z\|^2 \, \|q\|},$$

and beyond them:

$$\varepsilon < \frac{L - \|p_0\|}{\|q\|}.$$

- $Q \in \mathbb{P}_{n,L}$ and
- $|Q'(z)| > |p'_0(z)|$,

contradicting the assumption.

On the other hand from Chebyshev's observations ([Cheb54] and [Cheb59]) we know the only two functions which n-times reach L as their maximal value on $[a,b]$[59]— it is $\pm T_n$. Transformed for our interval we get the solution f_0 as

$$f_0(x) = \pm L \cos n \arccos \left(\frac{2x - a - b}{b - a} \right) \tag{3.31}$$

with the derivative

$$f'_0(x) = \pm \frac{nL}{\sqrt{(z-a)(b-z)}} \sin n \arccos \left(\frac{2x - a - b}{b - a} \right). \tag{3.32}$$

The function f_0 is the only one which satisfies the necessary condition. With this result Markov could prove an interesting property of polynomials of nth degree which we want to denote:

Theorem 3.3 *Let* $P : [a,b] \to \mathbb{R}$ *be a polynomial of degree n. Then we have for any $x \in [a,b]$:*

$$|P'(x)| \le \frac{n}{\sqrt{(z-a)(b-z)}} \|P\|. \tag{3.33}$$

As a separate theorem (3.33) was published firstly by S. N. Bernstein (Corollary 3 в from [Bern12/2]).

Since $(z-a)(b-z)$ reaches its minimum at $z = (a+b)/2$ the above-mentioned result shows the validity of the following inequality, now named after A. Markov:

---

[59] Indeed, Markov's alternation condition gives n equations for the $n+1$ coefficients of the polynomial and because of the restriction to polynomials of one norm we get the condition we seek.

Formally we show exactly:

The problem (A) posed by A. Markov is equivalent to the following problem (B) :

$$\text{Minimize} \quad \|p\|, \quad p \in \mathbb{P}_n,$$
$$\text{under} \quad p'(x) = 1 \quad \text{for a fixed} \quad x \in [a,b].$$

Proof. Let p be a solution of (A) with $p'(x) = M$. Then $\tilde{p} := \frac{p}{M}$ satisfies the side-condition of (B). Now assume that for $\hat{p}$ there also holds $\hat{p}'(x) = 1$. Additionally let $K := \|\hat{p}\| < \|\tilde{p}\| = \frac{L}{M}$. $\frac{L\hat{p}}{K}$ surely satisfies the side-condition of (A). Thus, $\frac{L\hat{p}}{K}'(x) = \frac{L}{K} > M$, so p cannot be a solution of (A). The other direction can be shown in the same way.

It is easier to analyse this minimization problem. We clearly see the formulation of the side-condition as a linear equation of the coefficients.

Theorem 3.4 *For all polynomials $p \in \mathbb{P}_n$ there holds:*

$$\|p'\| \leq \frac{2n^2}{b-a}\|p\|, \quad p \in \mathbb{P}_n. \tag{3.34}$$

3.3.2.2 Markov Systems

In his work 'About Extremal Values of Integrals Connected with the Interpolation Problem' [MarA98], A. Markov managed to solve the generalized problem of Korkin and Zolotarev.

Firstly he defined the today so-called 'Markov systems.' These are systems of functions $\lambda_1, \lambda_2, \ldots, \lambda_n, \ldots$, satisfying for each n

$$\lambda_1(z) > 0, \tag{3.35}$$

$$\begin{vmatrix} \lambda_1(z) & \lambda_1'(z) \\ \lambda_2(z) & \lambda_2'(z) \end{vmatrix} > 0,$$

$$\vdots$$

$$\begin{vmatrix} \lambda_1(z) & \cdots & \lambda_1^{(n)}(z) \\ \vdots & \ddots & \vdots \\ \lambda_{n+1}(z) & \cdots & \lambda_{n+1}^{(n)}(z) \end{vmatrix} > 0.$$

From this definition it directly follows that Vandermond's determinant evaluated with this function is regular. And so there holds the property which is used up to now to avoid the n-times differentiability Markov had demanded.

He could solve Korkin's and Zolotarev's problem, the determination of the polynomial least deviating from zero according to the L_1-norm, for this system of functions. To be precise, he proved

Theorem 3.5 *Let $\lambda_1, \lambda_2, \ldots, \lambda_n, \ldots$, be a system of functions on $[a, b]$ satisfying (3.35). Let Λ_k be the set of real linear combinations of the $\{\lambda_1, \ldots, \lambda_k\}$ and $f \notin \Lambda_k$ a function.*[60]

Then:

1. The functions

$$F_k(x) := \sum_{i=1}^{k} c_i \lambda_i(x) \tag{3.36}$$

satisfying

$$\int_a^b |\lambda_{k+1}(x) - F_k(x)|\, dx = \min_{p \in \Lambda_k} \int_a^b |\lambda_{k+1}(x) - p(x)|\, dx$$

are uniquely determined.

---

[60] Markov did not make any special assumptions for f. The following theorem holds for all L_1-integrable functions.

2. *The error of approximation of* f *by functions* $F_k(.)$ *is*

$$E_k(f, L_1) = \int_a^b |f(x) \operatorname{sign}[F_k(x)]| \, dx.$$

A. Markov used only algebraic methods to prove this theorem, he mainly made considerations about zeros and solved linear systems of equations. We could have expected this, since he did not discuss the properties f should satisfy.

3.3.2.3 The Lecturer Andrey Markov

As we have already mentioned in the section about his biography, Andrey Andreevich Markov worked 26 years at the university (from 1880 until 1905). In contrast to Chebyshev, A. Markov also lectured on the foundations of analysis, namely 'Introduction to analysis' (1882–1900) and 'Differential calculus' (1880–1882 and 1899–1905).

Traditionally 'Introduction to analysis' [MarA88] introduced concrete calculations with numbers and limits, therefore there were nearly no definitions.

The only available notes of the lectures on differential calculus [MarAoJ] are not dated, but presumably they describe the later cycle because the lectures from 1880–82 had been announced as repetitions—the main course had been given by Sochocki.

So if we assume that the lectures were given just before the end of the 19th century, we wonder if the definition of continuity was not exact:

> "A function $f(x)$ is called continuous for a value $x = a$, if $f(x)$ is approaching the limit $f(a)$ if the variable x is approaching the limit a."[61]

Then he restricted the definition arguing that it is necessary to approach a from both sides to guarantee that the limit is unique. The formulation 'approach' might have not included all sequences $x_n \to a$.

If we compare it with the definition Zolotarev had used about 30 years before Markov, we do not see a fundamental difference. There was no reasonable discussion of the concept of a sequence, and so the concept 'approach' remained slightly inconsistent.

Of course we can apologize for this—on his behalf he lectured on differential calculus when there had not been a complete clarification (1870), but these 30 years of intensive discussion among western mathematicians had passed Andrey Markov without obvious traces.

Another interesting fact is that Markov proved Weierstraß' theorem about maxima (continuous functions reach their maxima on closed intervals), but

---

[61] [MarAoJ, S. 16]: «Функция $f(x)$ называется непрерывной для значения $x = a$, если $f(x)$ приближается к пределу $f(a)$ при приближении переменного x к пределу a.».

did not mention the Riemannian and Weierstraß'ian 'monster,'[62] although they were famous at that time and very important for the discussion of mathematical concepts. Markov showed only examples with jumps and gaps to present examples for non-continuous and non-differentiable functions, respectively. Here we got again 'examples from practice.'

At the end of this chapter we will return here to a deep analysis A. Markov's 'Lectures about Functions the Least Deviating from Zero.'

3.3.3 Vladimir Andreevich Markov

Vladimir Andreevich Markov, Andrej Andreevich's youngest brother, was the second character within within St Petersburg Mathematical School who suffered a tragic fate. His early death prevented a hopeful mathematical career, similar to Zolotarev's fate.

3.3.3.1 About his Biography

Vladimir Andreevich Markov was born in St Petersburg May 8, 1871 as the youngest son of Andrey Grigorevich Markov and his second wife, Anna Iosifovna Markova.

Already like Andrey, Vladimir visited the fifth St Petersburg grammar school, and in 1888 he entered the physico-mathematical faculty of St Petersburg University. During his student's life the teachers who held the chair of pure mathematics were his brother Andrey Markov (Introduction to analysis, finite differences, probability theory), Posse (differential and integral calculus), Sochocki (integral calculus, higher algebra, number theory), Korkin (differential equations, variational calculus), Budaev (analytical and higher geometry), Ptashitski (descriptive geometry, elliptical functions), Selivanov (special courses about higher algebra and number theory) and Grave (geometrical applications of differential and integral calculus).

Already during his studies he caused a great stir: In 1892, as a student of the fourth year, he published the paper 'About Functions Least Deviating from Zero on a Given Interval' [MarV92] for which he had been awarded a prize on the occasion of the first congress of natural scientists and physicians. The positive report about it had been written by K. A. Posse,[63] who would support V. A. Markov's further career. We will discuss this work below.

After the end of his studies he remained at the university to prepare the job of a lecturer and began to write his master thesis about positive quadratic forms of three variables, stimulated by Korkin. He defended it in 1896.

---

[62] Riemann's "monster" is a function which is continuous for all irrational values and discontinuous elsewhere; Weierstraß' monster is a continuous, but nowhere differentiable function.

[63] Compare [Ser97, p. 43].

In the academic year 1895/96 he was repetitor (scientific assistant) at the St Petersburg institute for transportation,[64] where he lectured on analytical geometry. Besides this he taught at the fifth St Petersburg grammar school.

In the academic year 1896/97 he was not able to work, anymore. January 18, 1897, at the age of 26, Vladimir Andreevich Markov died from tuberculosis.

Only after his death was his master thesis published [MarV97].

3.3.3.2 A Student's Paper

In the above-mentioned work 'About Functions Least Deviating from Zero on a Given Interval' [MarV92] Vladimir Markov investigated the problem to determine the polynomial of degree n which deviates as little as possible from zero among all those polynomials whose coefficients satisfy a linear side-condition.

Such a kind of problem was not new—we already knew the problem of Zolotarev, who also proved some general theorem for some special side-conditions in [Zol68], and also A. Markov described a problem of this kind. One of the most remarkable aspects of V. Markov's monograph is the way he solved this problem. He presented a *complete theory* of the problem. This led to the fact that, rather than his central idea, some corollaries became famous, e. g., the exact border for the relation between the norm of the kth derivative of a polynomial and its norm.[65]

After this introduction we do not wonder that V. Markov proved another special alternation condition characterizing the solution of this problem.

3.3.3.3 Vladimir Markov's Problem

So V. A. Markov formulated the following problem:
Minimize $\|p\|$, *where*

$$p(x) = \sum_{i=0}^{n} a_i x^i \in \mathbb{P}_n, \quad x \in [a, b]$$

and its coefficients satisfy a linear side-condition

$$\alpha = \sum_{i=0}^{n} \alpha_i a_i$$

with given real numbers $\alpha, \alpha_0, \ldots, \alpha_n$.

Here Chebyshev's first case was generalized. An arbitrary coefficient is given, not absolutely the first.

Not exactly following V. Markov's terminology, we want to abbreviate and define:

---

[64] Russian name «институт инженеров путей сообщения»

[65] This Bernstein pointed out in his introduction to the German translation of the text [MarV16].

Definition 3.6 *Let* $\alpha_1, \ldots, \alpha_n \in \mathbb{R}$.
For $p \in \mathbb{P}_n$ *with* $p(x) = \sum_{i=0}^{n} a_i x^i$ *define*

$$\omega \; : \; \mathbb{P}_n \to \mathbb{R} \quad \text{with}$$
$$\omega(p) := \sum_{i=0}^{n} \alpha_i a_i. \tag{3.37}$$

For $\alpha \in \mathbb{R}$ *define*
$$\mathbb{P}_n^\alpha := \{p \in \mathbb{P}_n \mid \omega(p) = \alpha\}. \tag{3.38}$$

We remark that ω is linear.
Using these definitions we can now formulate the problem like this:

$$\min_{p \in \mathbb{P}_n^\alpha} \max_{x \in [a,b]} |p(x)|. \tag{3.39}$$

A solution of this problem is now called a minimal solution on $\mathbb{P}_n^\alpha$.

To solve this problem V. Markov proved a special alternation theorem. Preparing this he showed the following lemma which is a special case (and a real forerunner) of the Kolmogorov criterion:

Lemma 3.7 *Let* $p \in \mathbb{P}_n^\alpha$ *and* $x_1, \ldots, x_\mu$ *its deviation points.*
Then:
p is minimal solution on $\mathbb{P}_n^\alpha$

$$\Longleftrightarrow$$

There does not exist any polynomial $q \in \mathbb{P}_n$ *with*

$$\omega(q) = 0, \quad q(x_i)p(x_i) < 0 \quad \forall i = 1, \ldots, \mu. \tag{3.40}$$

Proof We set $L := \|p\|$ and choose $q \in \mathbb{P}_n$ with $\omega(q) = 0$
Furthermore define for $\varrho > 0$

$$p_1 := p + \varrho q. \tag{3.41}$$

With this we have because of the linearity of ω :

$$\omega(p_1) = \omega(p) - \varrho\omega(q) = \omega(p). \tag{3.42}$$

Now we want to assume that q satisfies (3.40). Then we can choose a ϱ so that $\|p_1\| < \|p\|$, as we now will show.

Since for all deviation points x_i, $i = 1, \ldots, \mu$ of p we have

$$p(x_i)q(x_i) < 0,$$

this remains valid in small neighbourhoods $U_\delta(x_i)$, $\delta > 0$ and of course for all elements of

$$U := \bigcup_{i=1}^{\mu} U_\delta(x_i).$$

Now choose

$$\varrho < \min \left\{ 1 \quad , \min_{x \in U} \left(\frac{-2p(x)q(x)}{\|q\|^2} \right) , \min_{x \notin U} \left(\frac{L^2 - [p(x)]^2}{|2p(x)q(x)| + [q(x)]^2} \right) \right\}. \quad (3.43)$$

Then we have on the whole interval $[a, b]$:

$$L^2 - [p_1(x)]^2 = L^2 - [p(x)]^2 - 2\varrho p(x)q(x) - \varrho^2 [q(x)]^2 \quad > \quad 0, \quad (3.44)$$

hence $\|p_1\| < \|p\|$. So, because of (3.42) p is not minimal.

To prove the other direction we firstly state that we can split every $r \in \mathbb{P}_n$ into

$$r := p + \sigma q$$

with $q \in \mathbb{P}_n$, since ω is linear.

If there does not exist any function in $\mathbb{P}_n$ satisfying (3.40), then there is a deviation point x_j of p with

$$p(x_j)q(x_j) \geq 0.$$

The equation analogous to (3.44) for r then leads to

$$L^2 - [r(x_j)]^2 < 0, \quad (3.45)$$

and so p is indeed a minimal solution on $\mathbb{P}_n^\alpha$.

3.3.3.4 An Alternation Theorem by V. A. Markov

The second means to solve V. Markov's problem was Lagrange's interpolation formula. At first we define

Definition 3.8 (Lagrange functions) *For $p \in \mathbb{N}$ and $x_1, \ldots, x_p \in \mathbb{R}$ define $F, F_l : \mathbb{R} \to \mathbb{R}$ with*

$$F(x) := \prod_{i=1}^{p} x - x_i$$

$$F_l(x) := \prod_{i=1, i \neq l}^{p} x - x_i \quad = \quad \frac{F(x)}{x - x_l}. \quad (3.46)$$

And now we can formulate the alternation theorem:

Theorem 3.9 (V. A. Markov's Alternation Theorem) *Let $p \in \mathbb{P}_n^\alpha$ with deviation points $x_1, \ldots, x_p$.*
Then:

p is a minimal solution on $\mathbb{P}_n^\alpha$

$$\Longleftrightarrow$$

$$1) \ \text{sign}\,(-1)\omega(F_1)p(x_1) = \text{sign}\,(-1)^2\omega(F_2)p(x_2) = \cdots \qquad (3.47)$$
$$= \text{sign}\,(-1)^p\omega(F_p)p(x_p).$$

2) *If* $p < n+1$, *then* $\forall R \in \mathbb{P}_{n-p} : \omega(FR) = 0$.

Proof With Lagrange's interpolation formula there holds for an arbitrary polynomial $q \in \mathbb{P}_n$ and an $x \in [a, b]$

$$q(x) = AF(x)R(x) + \sum_{i=1}^{p} \frac{q(x_i)}{F'(x_i)} F_i(x), \qquad (3.48)$$

where $A \in \mathbb{R}$ and $R \in \mathbb{P}_{n-p}$ for $p < n+1$; $R \equiv 0$ else.

The linearity of ω leads to

$$\omega(q) = A\omega(FR) + \sum_{i=1}^{p} \frac{q(x_i)}{F'(x_i)} \omega(F_i), \qquad (3.49)$$

having in mind that

$$F'(x_i) = (-1)^{p-i} \prod_{l=1}^{i-1}(x_i - x_l) \prod_{l=i+1}^{p} (x_l - x_i),$$

then we can write (3.49) as

$$\omega(q) = A\omega(FR) + \sum_{i=1}^{p} \frac{q(x_i)}{|F'(x_i)|} (-1)^i\omega(F_i). \qquad (3.50)$$

If we assume that for $p < n+1$ there is a function $R \in \mathbb{P}_{n-p}$ with $\omega(FR) \neq 0$, then we can define

$$q(x_i) := -p(x_i), \quad \text{i. e.,} \quad q(x_i)p(x_i) < 0.$$

Hence the constant A of equation (3.49) can be chosen so that $\omega(q)$ will vanish. From the lemma we know that then p can't be minimal.

With this remark we can write (3.50) as

$$\omega(q) = \sum_{i=1}^{p} \frac{q(x_i)p(x_i)(-1)^i\omega(F_i)p(x_i)}{|F'(x_i)|[p(x_i)]^2}. \qquad (3.51)$$

But then the equations $\omega(q) = 0$ and the condition $q(x_i)p(x_i) < 0 \ \forall i = 1, \ldots, p$ can be valid together, only if not all expressions of the form $(-1)^i\omega(F_i)p(x_i)$ are equal in sign. Using the lemma the theorem is proved.

The proof of this theorem was more complicated than the direct proof of the general alternation theorem, so it is justified to judge that it was a historical incidence that it had not been proved before the end of the 19th century in St Petersburg. We can assume that the general alternation property had already been known.

After the proof of his special alternation theorem V. Markov stated some facts about the number of deviation points. They are complicated in general because without the unicity of the solution there may be fewer deviation points:

Lemma 3.10 *If there is more than one polynomial $p \in \mathbb{P}_n^\alpha$ solving (3.39), then among the solutions there is at least one which has not more than μ deviation points, where*

$$\mu = \frac{n+2}{2}, \quad \text{if } n \text{ is even,}$$

$$\mu = \frac{n+1}{2}, \quad \text{if } n \text{ is odd.}$$

Therefore he quickly turned to the special case

$$\omega(p) = p^{(k)}(z) \tag{3.52}$$

for a given z.

3.3.3.5 The Special Side-Condition $\omega(p) = p^{(k)}$

For this case he was able to show the unicity of the solution, because here the solution has only real roots, which is due to the characteristic equations (2.22).

But then again we have a sufficiently large number of deviation points and get a solution equal to the Chebyshev polynomial (neglecting a constant factor). It is

$$p(x) = \frac{\alpha}{T_n^{(k)}(z)} T_n(x).$$

Further considerations of exclusively algebraic nature, which use properties of zeros, led to the well-known inequality, which gave the exact border for the relation between the norm of the kth derivative of a polynomial and its norm and was later named after V. Markov.

Theorem 3.11 (Vladimir Markov's Inequality) *Let $p \in \mathbb{P}_n$ be a polynomial. Then:*

$$\|p^{(k)}\| \le \left(\frac{2}{b-a}\right)^k \frac{n^2(n^2-1^2)\dots(n^2-(k-1)^2)}{(2k-1)!!} \|p\|, \quad p \in \mathbb{P}_n. \tag{3.53}$$

3.4 Julian Karol Sochocki

Although Julian Karol Sochocki[66] did not publish any contribution to the approximation theory, he deserves being mentioned at this place because it was due to him that function theory, widely spread in Western Europe, was also recognized as a subject of research in St Petersburg. In some way Sochocki set a counterpart to the mainstream of the St Petersburg Mathematical School defined by Chebyshev and his work.

3.4.1 About his Biography

Julian Karol Sochocki was born February 5, 1842 in Warsaw as a son of the public servant Bazyli Sochocki.[67] From 1850 until 1860 he visited the physico-mathematical department of the grammar school of the Warsaw region. Immediately after this he registered at the physico-mathematical department of St Petersburg University.

His university education, however, was interrupted in March, 1861: February 27, 1861 several Polish nationalists were shot by the police in Warsaw whereupon some St Petersburg students manifested their solidarity by a requiem. As a counter-measure all students of Polish origin were held to account, among them Sochocki who indeed took part in the protests. He could avert the probable punishment only by returning to Warsaw because of "domestic duties." During the 1863/64 riots in Poland he rendered logistic assistance to the revolters, but remained unknown to the the the powers. At that time he could only deal with mathematics by self-studies and returned to St Petersburg only in 1865 as an auditor.

In the following year he presented his candidate thesis and was allowed to stay at the university to prepare for a professor's appointment. In 1868 he defended the master thesis "Theory of residuals with some applications" (теория вычетов с некоторыми приложениями), for which his opponents were Somov and Chebyshev.

An important result of Sochocki's thesis was a theorem about essential singularities which he proved independently from F. Casorati (1835–1890) and which is known now as the Casorati–Weierstraß theorem—Weierstraß' proved it by other means, but only in 1876.

Probably Sochocki was already encouraged in this subject in Warsaw. In 1862 the Polish university had been founded there and its physico-

---

[66] Although Sochocki is a Pole, mostly the Russian form of his name, Julian Vasilevich Sochotski, is used. This can be justified, since Poland's Eastern part politically belonged to the Russian Empire at that time and so the Russian spelling was the official one. We want to allow for his origin and use the Polish form as used, e. g. in [Erm98/1].

[67] The biographical data were taken from [Erm96] and were supplemented by some information from [Erm98/1].

mathematical department was ruled by Augustyn Frączkiewicz (1796–1883), a pupil of Cauchy.

The possibilities to deal with function theory at the St Petersburg University were not good because Chebyshev himself behaved with indifference towards it and did not use its results.

In 1873 Sochocki defended his doctoral thesis About definite integrals and functions used with series expansions" (Об определенных интегралах и функциях, употреблаемых при разложении в ряды) from the opponents Chebyshev and Korkin.

In this work he proved some formulae for the calculation of limits of integrals of Cauchy type, often named after Josif Plemelj (1873–1967), who did not prove them until 1908.[68]

In the doctoral thesis he also showed a connection between function theory, continuous fractions, orthogonal polynomials and the approximate computation of integrals of Cauchy. These considerations had a wider basis, but less concrete results than those by Posse, as he later would present in his master thesis "About functions similar to Legendre's" [Pos73].

From 1868 Sochocki lectured at the St Petersburg university, first as "privat-dotsent," one year later as an ordinary lecturer, from 1873 as an extraordinary professor, from 1882 as an ordinary and finally since 1893 as a merited professor. Not before 1923 did he left the university. In parallel he directed the chair of mathematics at the institute of civil engineering.

In 1894 he was elected corresponding member of the Cracow Academy of Sciences.

The subjects of his lectures[69] were "Theory of continuous fractions" (1868–71), "Theory of functions of imaginary quantities" (1868–71), "Analytical geometry" (1871–73), "Higher algebra" (from 1871), "Differential and integral calculus" (1873–82), "Introduction to analysis" (1876–80), "Integral calculus" (from 1882)[70] and "Number theory" (1882–92).

He also did represent function theory in his lectures. The integral calculus lectured over 40 years included its basics and elliptic functions, which had been taught separately before him (by Somov, Bessel and Zolotarev). His lectures on higher algebra were published in two volumes.

The end of his life was tragic because of the hard times after the 1917 revolution. He spent his last two years in the old peoples' home of the house of Leningrad scientists and died there on December 14, 1927.

---

[68] There are even some other theorems proved first by Sochocki, but named after other mathematicians (comp. [Erm96, p. 361]).

[69] Comp. [VorSPb].

[70] This lecture had been presented with several different names, e. g., "Theory of definite integrals," "Theory of multiple integrals" and so on.

3.4.2 Chebyshev's Supplement

Sochocki's work and his pedagogical job were apparently not influenced by Chebyshev, although he had been an opponent at the defence both of Sochocki's master's and doctoral thesis.

Also his subjects had nothing in common with those of Chebyshev. Obviously Sochocki had been influenced mainly by the Western European Cauchy school in Warsaw.

This is an indirect proof for the strength of Chebyshev who at times clearly rejected the "philosophizers" and even attacked Cauchy,[71] but recognized the significance of his theory and supported the first St Petersburg representative of that mathematical direction. Surely Chebyshev knew about the harmful effect of any unnecessary constriction of mathematics.

Thus, Sochocki did not differ from other representatives of the St Petersburg Mathematical School in his adoration of the mathematical idol Chebyshev. At the session of the St Petersburg Mathematical Society on January 14, 1895 he spoke out on his work:

> "We study them under all aspects; we will pass them to the next generation with necessary supplements and condign clarifications as a pledge of the further independent development of the mathematical sciences in Russia and as a pledge of the unchanged relatedness to the scientist and compatriot whose name will live as long as the science will do."[72]

As we will see, Sochocki found an associate in Posse, who also tried to follow newer developments even from outside Russia.

3.5 Konstantin Aleksandrovich Posse

Konstantin Aleksandrovich Posse played an important role as teacher and lecturer at St Petersburg University. He lectured more than 40 years with Chebyshev, Korkin and A. A. Markov and especially cared about the students' education in calculus. His textbooks on differential calculus (e. g., [Pos03]) were published several times.

---

[71] Comp. [Erm87]. In the notes about Chebyshev's "Probability Theory" commented there we find some critical remarks about statements attributed to Cauchy. In addition Chebyshev's biographers always emphasize that the personal relationship between Chebyshev and Cauchy was dry.

[72] Protocols of the St Petersburg Mathematical Society, 1899, cited after [Erm96, p. 364]: «Мы изучаем их всесторонне; снабдив необходимыми дополнениями и надлежащим освещением, своевременно мы передадим их следующему поколению в залог дальнейшего, самостоятельного развития математических наук в России и в залог неизменной признательности ученому-соотечесвеннику, имя которого будет жить столько, сколько будет жить сама наука.»

3.5.1 His Biography

Konstantin Aleksandrovich Posse was born September 29, 1847 in St Petersburg as the son of the railway engineer, Aleksandr Fedorovich Posse, who was Swedish by descent, and his wife Elizaveta Jakovlevna Posse.[73]

At first he was educated at home and in 1860 he entered the fourth class of the second St Petersburg grammar school, a school whose teachers were supported by university as well and so enjoyed a good reputation. In 1864 he finished grammar school with the gold medal.

In the same year he was allowed to visit the physico-mathematical faculty only as an auditor but because of his youth, but after one year he became a full student when the university took into account his former efforts. Obvious parallels with the biography of his fellow student Egor Ivanovich Zolotarev, who came to university one year earlier, can be seen, and we again want to emphasize that these had been very unusual occurences.

His academic teachers were the same as those of Zolotarev—Somov, Chebyshev, and Korkin.

In 1868 K. A. Posse finished his studies with a candidate dissertation about Euler integrals of first and second degree. In 1870, before the opponents Chebyshev and Sochocki, he passed his master examinations: In 1873 before the opponents Korkin, Somov and Zolotarev, he defended his master dissertation 'About Functions Similar to Legendre's' [Pos73], which we will talk about later.

Posse's doctoral dissertation 'About θ-functions of two variables and a problem of Jacobi' dealt with the theory of elliptic functions, especially with integrals of the form

$$\int f\left(x, \sqrt{R(x)}\right) \, dx, \quad R \text{ a polynomial,}$$

and was defended in 1882. Here his opponents were Korkin and Sochocki. In this work Posse discussed and unified approaches of Rosenhain, Weierstraß and Riemann.

His university career began in 1874, one year after the defence of the master dissertation. Until 1883 he read analytical geometry, first as a 'privat-dotsent', in 1880/81 as an ordinary lecturer together with an introduction to analysis. In 1883 he became an extraordinary professor and started to lecture on calculus, often named as 'Applications of calculus to analysis and geometry.' With different names these lectures were given by Posse until 1899,[74] when he left the St Petersburg University after becoming a merited professor. Again we

---

[73] Biographical data were taken from [Ser97]. In this section political and social developments are described more carefully, because the biographer Sergeev was (allowed to be) more objective than the other authors who wrote their contributions during Soviet times.

[74] Comp. [VorSPb].

see that usually one lecture was given only by one certain lecturer. The 1883's change was connected with Chebyshev's departure from the university.

After his departure Posse was appointed honorary member of the university. In the years 1919–1921 he again taught at the university under severe conditions.

In 1894/95 Posse spent one year abroad to be cured of a slight medical complaint. To protect him from negative financial consequences, the university declared this trip as an official scientific trip.

He also lectured at other St Petersburg institutes: From 1871 until 1882 and from 1890 until his retirement at the transportation institute, at the womens' university[75] since its foundation in 1876 until 1886 and from 1900 until its merger with the Petrograde University in 1919, at the technological institute from 1891 until 1894 and at the electrotechnical institute from 1899 until 1905.

The subject of all these lectures was calculus, the mathematical basics for prospective engineers. Thus, over a 30 years time frame a large number of Russia's best engineers, who usually studied at the capitol's institutes, learned the theoretical foundations of their subjects from Posse.

Posse also took an active part in organizations supporting mathematics using his excellent reputation as a teacher. He was one of the founding members of the St Petersburg Mathematical Society (1890). In the initial years he was its speaker together with Sochocki, Bobylev and V. I. Shiff. In 1908 Posse became a member of the Russian subcommission for the reform of mathematical education; in 1914 he was appointed Russia's representative in a similar international commission. On behalf of this commission he was engaged in questions of education of mathematical analysis at schools and of the function of mathematics at technical universities. About these activities he reported at a conference in Paris which took place on the eve of the First World War. Among its participants there were É. Borel, J. Hadamard, G. Darboux and H. Lebesgue. In his reports Posse always rejected the claim of a decline of the theoretical level of technical universities.

In 1913 he undertook, on behalf of the ministry of education, reviewing of newly published mathematics textbooks for schools and universities. In 1915 (after the death von N. Ya. Sonin he became speaker of the commission for mathematical education within the committee for the reformation of school education at the ministry of education.

After the revolution of February 1917 Posse became a member of the "association for the development and promulgation of positive sciences," which was meant to be a supplementary organization to the academy of sciences, to deal with publishing, education and some additional scientific jobs. Among the other members were V. A. Steklov and the writer M. Gorkij. Like most Russian intellectuals Posse took the new found freedom with enthusiasm. The

---

[75] They were also known as 'Bestuzhev-courses' after their patron K. N. Bestuzhev-Rjumin.

activities of the association, however, ended with the revolution of October 1917.

Posse also was an active member and patron of several organizations, within and beyond the sciences. As a member of the fund-raising society of the women's university, which also was a steering committee of this university, he organized syllabuses and the build-up of a library.

The extraordinarily talented pianist K. A. Posse was an active member of the St Petersburg society for chamber music consisting of about 60 members most of whom were professional musicians; among them were P. I. Tchaikovsky, N. A. Rimski-Korsakov, A.G. Rubinstein and others. This society organized concerts, where Posse himself played. But he mainly took over "delicate jobs of fund-raising."[76] Posse had a similar role with the committee for the promotion of the literary heritage, which supported emerging writers. The committee was founded by an initiative of A. K. Tolstoy, I. S. Turgenev and N. G. Chernyshevski.

For the middle-class liberal Posse, the October revolution marked a bitter turning point.

Because of his illness in 1921 he gave up his lecturer job at the age of 74. His hopeless financial situation forced him to spend his eventide in a badly supplied old peoples home.[77]

Konstantin Aleksandrovich Posse died there on August 24, 1928.

3.5.2 The Scientific Work of K. A. Posse

In some way Posse's work represented a bridge between the function theoretical work of Somov, Sochocki and Zolotarev and the core problems of the St Petersburg Mathematical School, the extremal problems.

3.5.2.1 Orthogonal Polynomials

Posse's master thesis 'About Functions Similar to Legendre's' from 1873 [Pos73] tied in with Chebyshev's former contributions on this subject (e. g., [Cheb55/2] and [Cheb70]). The expression similar to Legendre's obviously can be understood as "orthogonal."

Posse regarded orthogonal polynomials as finite fractions of the expansion of the integral

$$\int_a^b \frac{f(z)}{x - z}\, dz$$

into a continuous fraction.

---

[76] As his biographer Sergeev formulated.

[77] Actually he spent his last days in a so-called "Internat" «интернат», a house for the poor, but only elder people were there at that time. By coincidence, Sochocki lived in the same house for the rest of his life.

A well-known result is that the Legendre polynomials P_n satisfy the following property:

They differ from the nth partial fraction of the expansion of

$$\frac{1}{2} \log \frac{x+1}{x-1}$$

only by a constant, and this expression is equal to the integral expression

$$\int_{-1}^{1} \frac{1}{x-z}\, dz.$$

So the integral expression investigated by Posse is a possible generalization of this expression.

Posse now showed that this case includes all orthogonal polynomials discussed by Chebyshev, if f is replaced by the respective weight function. And so Posse was able to prove Chebyshev's results in one step.

It is worth mentioning that sometimes Posse did use Cauchy's integral formula

$$f(z) = \frac{1}{2\pi i} \int_{\gamma} f(x)x - z\, dz$$

for a closed path γ. He himself emphasized that the impulse had been given by Sochocki.[78]

He deepened his results in several other contributions to this subject. In 1886 he summarized his results and cited well-known theorems of other authors in the monograph ,,Sur quelques applications des fractions continues algébriques" [Pos86].

3.5.2.2 Other Works

Posse's scientific heritage is small, if we compare it with the work of other members of the St Petersburg Mathematical School. His main contributions dealt with applications of expansions into continuous fractions, where he also investigated several problems about minima and even moments.

Besides them there are papers with a clear accordance to his pedagogical work. In 1895 the work 'About the transcendence of e and π' was published, where he presented Lindemann's results in a way which was easier to read.

He also wrote about applied subjects coming from his jobs at several engineering schools.

However, his main work was the different textbooks on calculus edited by him. Usually they were re-edited, but Posse always tried to take into account all new directions. From 1891 until 1939 twelve editions were published and some of them even were translated, as into German in 1923. There was no textbook on calculus which was spread wider in Russia than these editions.

---

[78] [Pos73, p. 3]: "I was led this way to the sought expansions by Sochocki's work 'theory of residuals' («На такой путь получения искомых разложений я был наведен сочинением г. Сохоцкого ,,Теория интегральных вычетов".»).

3.5.2.3 Posse's Basic Concepts

We see regarding his scientific work that Posse had no prejudices about new theories from Western Europe, like function theory. He was also interested in the foundations of analysis, which was of course a logical consequence of his pedagogical engagement.

In the non-published notes of a lecture of 1891 [Pos91] he introduced the concept of a function as another name for a dependent variable, in principle like Euler did. Analogously he classified functions into explicit and implicit ones and into algebraic and transcendental ones.

His definition of continuity was still a little awkward using the concept of infinitely small quantities, but he already used the modern $\varepsilon - \delta$-formulation:

"A function $f(x)$ is called continuous for a given value x, if the difference $f(x+h) - f(x)$ becomes infinitely small with an infinitely small h, in other words, if there exists a sufficiently small number α, so that for all values h, whose absolute values are smaller than α, or lying between $-\alpha$ and $+\alpha$, the absolute value $f(x+h) - f(x)$ is smaller than a given number ε, which can be set arbitrarily small."[79]

Thus, already in 1891 there was a precise definition of continuity in a Russian lecture.

In the printed version of these lectures of 1903 the complicated formulations were simplified and continuity was defined by limits of sequences and was also precise according to the first formulation.

But nevertheless differentiability was introduced in a more old-fashioned way:

"[...] in general the limit

$$\lim \left[\frac{f(x+h) - f(x)}{h} \right]_{h=0}$$

is a defined number, but in special cases, at some special values x, this limit might not exist."[80]

---

[79] [Pos91, p.15]: «Функция $f(x)$ называется непрерывной для данного значения x, если при h бесконечно малом и приращение функции $f(x+h) - f(x)$ будет бесконечно мало, иными словами, если существует такое, достаточно малое число α, чтобы для всех значений h численно меньших α, то есть лежащих в пределах $-\alpha$ и $+\alpha$, численное (абсолютное значение) разности $f(x+h) - f(x)$ было меньше заранее заданного числа ε, как бы мало ε ни было».

[80] [Pos03, p. 38]: «[...] вообще

$$\text{пред.} \left[\frac{f(x+h) - f(x)}{h} \right]_{h=0}$$

определенное число, а в частных случаях, при некоторых частных значениях x, предел этот может и не существовать.»

Why he spoke about special cases became clearer in the following remark:

"Nowadays some continuous functions are defined, which do not have a derivative <u>for any value of the independent</u> <u>variable</u>, but these functions do not have any applications, neither theoretical, nor practical, and we won't deal with them."[81]

Then he pointed out that the sense of these functions is the insight that continuity does not imply differentiability.

We see that Posse did discuss the foundations of analysis and went much farther than his predecessors. It is obvious why he was not interested in 'monsters.' Indeed there were no applications of them at that time—their most important sense had been mentioned by Posse!, and secondly maybe he did not want to overtax his students and readers, mostly engineers, with theoretical niceties.

However there remains a drop of irony, because even Posse emphasized the relation to applications as an important property of a mathematical *concept*.

Thus, in this sense Posse was also a typical representative of the St Petersburg Mathematical School.

3.6 A. A. Markov's Lectures 'About Functions Deviating the Least Possible from Zero"

A. A. Markov did not restrict his work to special problems. His lectures about functions deviating the least possible from zero (published 1906 [MarA06] and re-published in [MarA48]) contained the most important contributions of St Petersburg mathematicians to approximation theory.

Among the references of [MarA06] we only find the above-discussed St Petersburg contributions by Chebyshev, V. A. Markov and A. A. Markov himself and a note about Blichfeldt's remark.[82] Hence it is possible that Markov did not know both Borel's and Kirchberger's work when he wrote down his lectures. A. A. Markov's biographer, S. Ja. Grodzenskij [Gro87] did not give plausible hints[83] about when the lectures were given, but it is possible to find some evidence.

---

[81] [Pos03, p. 38]: «<u>Замечание:</u> В настоящее время указаны некоторые непрерывные функции, которые не имеют производной <u>ни при каком</u> <u>значении независимого переменного</u>, но эти функции никаких, ни теоретических, ни практических приложений не имеют, и мы ими заниматься не будем.»

[82] See section 4.2.

[83] Grodzenski stated that the lectures were given in 1906 at the university, but this information cannot be taken from the university calendar [VorSPb]. On the other hand this fact would be hard to believe, since the subjects of the lectures did not substantially change between 1869 and 1906.

3.6.1 Circumstances and Significance

It is clear that the lectures could not be given with this extent before 1892, since they contain Vladimir Markov's [MarV92] contributions. The citation of Blichfeldt's note from 1901 [Bli01] makes probable that they took place first in the 20th century. They did not belong to the official university program which can be shown by the university calendar from 1890 to 1906 [VorSPb].

So there remain two assumptions, where and why they were given. Firstly we remember that in St Petersburg it was not unusual to offer private courses and lectures to elder students, especially A. Markov's academic teachers A. N. Korkin and E. I. Zolotarev did so.[84] On the other hand the form of the lithograph gives a hint. The lectures existed together with two other lectures, the 'Notes about Mathematics' by A. N. Krylov (Записки по математике) and another lecture by A. A. Markov "About Continuous Fractions" (О непрерывных дробях). The lithograph was named 'Free Faculty' (Вольный факультет).

The free faculty was a voluntary offer of lecturers of St Petersburg university and other institutes during the 1905/1906 student's riots, which made it impossible to teach normally.[85]. The voluntary lectures were given at a private grammar school and the mathematical offer was organized by N. M. Gyunter who was a private lecturer of St Petersburg University at that time.

Evidence for A. Markov's engagement can be found in Jushkevich's book [Jush68] in the section about lectures of A. N. Krylov. Jushkevich writes: "[...] besides he himself [Gyunter] and Krylov [there] also lectured A. A. Markov, S. E. Savich and P. A. Shiff."[86]

Markov's biographer S. Ya. Grodzenski states that the lectures were given only once, which is probable because special lectures like these usually were not a part of the official calendar as we argued before.

3.6.2 The Posing of the General Problem

The general problem of the 'theory of functions least deviating from zero' was formulated as follows:

---

[84] Compare also [Akh48, p. 9].

[85] Also in other regions the studies were made more difficult. In a letter from February 19, 1906 to Hilbert [EgoHil] the Moscow professor D. F. Egorov explained his request to support his student Mikhail Kovalevski with the registration at Göttingen with the words: "I am attached to the interests of Mr Kovalevski because at present it is impossible to study in Russia and it would be a pity if this eager young man were condemned to idleness" („Die Interessen Herrn Kowalewskys liegen mir am Herzen, da es augenblicklich ganz und gar unmöglich ist, in Russland zu studieren und es jammerschade wäre, wenn der eifrige junge Mann zum Nichtstun verurteilt wäre.")

[86] [Jush68, S. 481]: «[...] помимо его самого и Крылова лекции читали А. А. Марков, С. Е. Савич и П. А. Шифф.»

Let a function
$$f : D \times \mathbb{R}^n \to \mathbb{R},$$
be given with variables $(x_1, \ldots, x_m)$ from the closed bounded domain (Markov named it 'finite') $\Omega \subset \mathbb{R}^m$ and parametres $(p_1, \ldots, p_n) \in \mathbb{R}^n$. Furthermore we assume that the partial derivatives
$$\frac{\partial f}{\partial p_i}$$
are continuous functions in Ω for all $i = 1, \ldots, n$ and that for all $i, j = 1, \ldots, n$ the second partial derivatives
$$\frac{\partial^2 f}{\partial p_i p_j}$$
are bounded in Ω.

These were the same assumptions which were needed by Chebyshev to prove Theorem 2.1 from page 38, but Markov formulated them more precisely (continuity of the partial derivatives) and the scope of them is expanded to functions of more than one variable. Thus, now to minimize was the quantity

$$L := L(p_1, \ldots, p_n) := \max_{(x_1, \ldots, x_m) \in \Omega} |f(x_1, \ldots, x_m, p_1, \ldots, p_n)| \qquad (3.54)$$

for all parametres $(p_1, \ldots, p_n) \subset \mathbb{R}^n$.

Markov remarked that because of one theorem of Weierstraß the quantity L exists for all parametres. But there was no remark about the existence of the minimum

$$E_n(f) := \min_{(p_1, \ldots, p_n) \in \mathbb{R}^n} L(p_1, \ldots, p_n),$$

neither at this place, nor later.

3.6.3 General Results

In the following sections Markov proved Chebyshev's theorem (Theorem 2.1) for the above-described generalized case.

As a means to that end there was a proved theorem from linear algebra, already used by Chebyshev, but proved only for the special linear system of equations (2.31), namely the fact that from the only trivially solvability of

$$Ax = 0$$

there follows the solvability of

$$Ax = b$$

for all b.

After that Markov left the general analysis of the problem, because Chebyshev's theorem only gives a necessary but not sufficient condition for the parameters of the function of least deviation from zero, as he argued. He wrote:

"But that it definitely deviates the least possible from zero with these parameter values, this fact has to be proved for any single case separately."[87]

This caused the discussion of special cases following now.

3.6.4 Polynomial Approximation

Markov dedicated the fourth section of his lectures to the most important characterizing theorem for the best Chebyshev approximation of continuous real-valued functions by polynomials of given degree, the alternation theorem. Now the functions which should be approximated were no longer defined on a higher-dimensional domain, but only on a closed interval.

As explained before, in [Cheb59] Chebyshev had only proved a necessary criterion for the number of deviation points and had not mentioned their alternating property. However, presumably the alternation of the signs of the deviation points was not an unknown fact, since A. Markov himself and his brother V. Markov proved it for some special cases ([MarA90] and [MarV92]) that we discussed before. Additionally A. Markov cited Blichfeldt's note [Bli01] among the references, which perhaps gave the last impulse to prove the theorem. Markov's method to prove the alternation theorem was elementary and looks like a translation of Borel's proof from [Bor05]. Nevertheless A. Markov probably himself found the proof, since at first it can be assumed that the date of the drawing up of the lectures was not later than the edition of Borel's 1905 lectures and secondly there is an important detail where Markov's proof differed from Borel's proof.[88].

In fact, Markov did not use topological arguments as did Borel, who mentioned that the set of deviation points must be closed and so assumed only a finite number of deviation points, which was not necessary in Borel's proof. Therefore A. Markov was not able to prove the alternation theorem completely. This improper assumption might have been caused by the fact that Markov had in mind only polynomials least deviating from zero. Indeed in the cases discussed by Zolotarev [Zol68], himself [MarA90] and his brother Vladimir [MarV92] there were only a finite number of deviation points.

With the help of the alternation theorem A. Markov could prove the unicity of the best approximating polynomial. But he did not prove its existence.

3.6.5 Extremal Problems: Polynomials Deviating the Least Possible from Zero

In the next sections Markov discussed the already well-known special cases ruling the St Petersburg Mathematical School: the determination of polynomials least deviating from zero under certain side-conditions for their coefficients.

---

[87] [MarA48, S. 252]: «Но что при этих значениях параметров она действительно наименее уклоняется от нуля, – это в каждом отдельном случае должно быть доказано особо.»

[88] See [Stef94] for a more detailed analysis of Markov's and Borel's proofs.

Markov solved Chebyshev's 'first case', which led to the determination of $T_n(x)$ as the polynomial of degree n of least deviation from zero with given first coefficient, with the equations

$$(f(x))^2 - L^2 = 0, \quad x \in [a,b] \quad \text{and}$$
$$f'(x) = 0, \quad x \in (a,b)$$

in the deviation points, as in [MarA84] and already [Cheb54].

Zolotarev's problem (comp. [Zol68] and [Zol77/1]) was solved only for an example because of the technical difficulties. Markov even called Zolotarev's methods unsuitable for practical purposes because of their difficulty.[89]

Other subjects of the lectures have also been discussed here, such as

1. approximation in $C\left(\frac{1}{\sqrt{\Psi}}\right)$ where Ψ is a polynomial (see [MarA84]),
2. 'Mendeleev's problem' of determining a smallest higher boundary for the derivative of a polynomial with a given norm [MarA90] and
3. Vladimir Markov's problem [MarV92] of determining the polynomial of least deviation from zero with coefficients satisfying a linear equation.

Besides this A. Markov discussed two other subjects, a modern analysis of Poncelet's problem and another approximation problem with side-conditions.

3.6.6 An Interpolatory Side-Condition

The ninth section was dedicated to the following problem. At first we set for a polynomial $p \in \mathbb{P}_n$ and a positive polynomial $\Psi \in \mathbb{P}_{2n}$:

$$f := \frac{p}{\sqrt{\Psi}}. \tag{3.55}$$

For a $\xi \in \mathbb{R} \setminus [-1,1]$ let

$$f(\xi) := h.$$

The problem is that of determining the function f which is, among all functions satisfying these side-conditions, the one which deviates as little as possible from zero on the interval $[-1,1]$.

In [Cheb80] Chebyshev discussed the same problem for the trigonometric polynomial

$$F(\varphi) := A_0 + \sum_{i=!}^{n} A_i \cos i\varphi + \sum_{i=1}^{n} B_i \sin i\varphi, \phi \in [-\phi_0, \phi_0].$$

With the transformation

---

[89] [MarA48, p. 264]: «Но решение Золотарева, основанное на применения эллиптических функций, слишком сложно для того, чтобы можно было им пользоватся на практике.»

$$x = \frac{\tan \frac{1}{2}\phi}{\tan \frac{1}{2}\phi_0}$$

we get

$$F(\phi) = f(x) = \frac{\sum\limits_{i=1}^{2n} P_i x^i}{\left(1 + x^2 \tan^2 \frac{\phi_0}{2}\right)^n}.$$

The problem was solved by factorizing numerator and denominator, using some equations for the coefficients determined in previous paragraphs of the lectures and applying the alternation condition.

3.6.7 The Main Subject of A. Markov's Lectures

P. K. Suetin writes about A. Markov's lectures:[90] "The formulations are of a complete modern character and are even nowadays repeated in monographs and articles about approximation theory. And besides this there is no doubt that some formulations belong to A. A. Markov and cannot be found in the work of other mathematicians."

However, we should analyse this opinion. Indeed the lectures convince in their attempt to arrange the different Russian contributions to approximation theory in a systematic way. But today we have to ask ourselves what the theoretical frame of these results was about.

Only Chebyshev's theorem (Theorem 2.1), slightly generalized by Markov, dealt with the general problem of approximating a continuous function by families of functions which are determined by a finite number of parameters.

Shortly after the proof of this theorem the problem was restricted to the approximation of continuous real-valued functions of one variable by polynomials. Here for the first time in the Russian literature the alternation theorem was mentioned and—with restrictions—proved. But also Markov avoided statements about the existence of solutions.

So we get the impression that both theorems have only been proved as means for the purpose solving the above-described special problems. He avoided discussion which theoretical results these theorems might imply.

Markov himself made clear what his aims were in his introductory remarks in the sixth section. After the determination of T_n as the solution of Chebyshev's first case he wrote to pass over to the other problems:

"Let us assume that a function $f(x)$ is given for all values $x \in [-h, h]$ by the convergent power series

---

[90] [Sue87, p. 210]: «Формулировки носят вполне современный характер и почти без изменений повторяются и в настоящее время в монографиях и статьях по теории приближения функций. В то же время некоторые формулировки несомненно принадлежат А. А. Маркову и не встречаются в предшествующих работах других математиков.»

$$f(x) = a_0 + a_1 x + a_2 x^2 + \cdots + a_{n-1} x^{n-1} + a_n x^n + a_{n+1} x^{n+1} + \cdots;$$

it is demanded to represent it by a polynomial of the form

$$q_{n-1} x^{n-1} + q_{n-2} x^{n-2} + \cdots + q_0,$$

so that the difference

$$f(x) - q_{n-1} x^{n-1} - q_{n-2} x^{n-2} - \cdots - q_0 \qquad (1)$$

least deviates from zero on the interval $[-h, h]$. After defining

$$q_0 = a_0 + \varepsilon_0, q_1 = a_1 + \varepsilon_1, \ldots, q_{n-1} = a_{n-1} + \varepsilon_{n-1},$$

we now will determine the corrections $\varepsilon_0, \varepsilon_1, \ldots, \varepsilon_{n-1}$ of the coefficients a_0, a_1, a_{n-1} under the condition that the function

$$\phi(x) = -\varepsilon_0 - \varepsilon_1 x - \cdots - \varepsilon_{n-1} x^{n-1} + a_n x^n + a_{n+1} x^{n+1} + \cdots \qquad (2)$$

on the interval $[-h, h]$ deviates the least possible from zero.[91] Finding the exact values of these for every given number n and for every function $f(x)$ satisfying the given condition is a very difficult problem, and we stand far away from a solution of it. Therefore we restrict ourselves to the determination of the first approximations for the corrections $\varepsilon_0, \varepsilon_1, \ldots, \varepsilon_{n-1}$ neglecting the terms $a_{n+1} x^{n+1} + a_{n+2} x^{n+2} + \ldots$ of expression (2) under the assumption that the sum of the neglected terms is sufficiently small."[92]

Thus, the circle was closed. We see that this was exactly the same problem Chebyshev had already posed in his first contribution to this theory, in „Théorie des mécanismes ... [Cheb54]" (comp. section 2.3) and which again was cited by Chebyshev as an important application of the results of „Sur les questions des minima ..." [Cheb59] in the introductory contribution [Cheb57].[93] Obviously the aim was not the determination of a best approximation for any continuous function, but only for a real-analytic function or an analytic approximative expression of certain functions. Chebyshev seemed to pose the problems for more general functions, but immediately restricted them to real-analytic functions.[94]

---

[91] We have refrained from citing the original text up to this point.

[92] [MarA48, p. 263]: «Нахождение точных величин поправок при всяком заданном n и для всякой функции $f(x)$, удовлетворяющей поставленным условиям, представляет весьма трудный вопрос, от разрешения которого мы очень далеки. Поэтому мы ограничимся разысканием первых приближенных значений поправок $\varepsilon_0, \varepsilon_1, \ldots, \varepsilon_{n-1}$, когда в выражении (2) отбросим все члены $a_{n+1} x^{n+1} + a_{n+2} x^{n+2} + \ldots$, предполагая, что сумма отброшенных членов ничтожно мала.»

[93] Compare section 2.4.8 of the present work.

[94] Compare ibid.

Also because of A. A. Markov's obvious missing knowledge of important Western European contributions (in this case especially Kirchberger and Borel) there was no attempt to lift the lectures to a more general level.

An especially remarkable aspect is the fact that the twenty-years-old Weierstraß' approximation theorem (from [Wei85]) was not mentioned by A. Markov. It is improbable that A. Markov did not know it, because, as we already know, there was an active interchange between mathematicians from St Petersburg and Berlin. Borel consequently discussed this, one of the basic theorems of approximation theory, at the beginning of the chapter „Représentation des fonctions continues par des séries de polynomes" [Bor05, S. 50-61] of his 1905 lectures and presented several proofs!

Markov's general theorems stand alone at the beginning of his lectures, isolated from concrete problems. Theory remained incomplete.

So mainly A. Markov's lectures remained a successful attempt to present the main results of the St Petersburg Mathematical School in a clear form.

In a certain way they mark the end of the first stage of the origin of approximation theory which was dominated to that time almost only by St Petersburg mathematicians.

3.7 Résumé: Practice by Algebraic Methods

With a small number of exceptions (Sochocki and Korkin) it was the work of Chebyshev which shaped the main subjects of the St Petersburg Mathematical School. We can show that the orientation to practical problems including applications within mathematics was the common sense of this school.

After Chebyshev some individual branches came into being and began to develop independently from each other: number theory (Voronoy, Ivanov), probability theory (A. Markov, Lyapunov), algebra (Grave), mathematical physics (Lyapunov) and others.

It is conspicuous that Chebyshev's inner circle did not share his enthusiasm for mechanism theory. This might have been caused by the difficulties of the problems which remained unsolved after Chebyshev's contributions and made new mathematical methods necessary.[95] Some exact analyses of Chebyshev's were only published after WWII.[96]

The spreading of Chebyshev's ideas also reached far away from St Petersburg. The St Petersburg Mathematical School, as the first mathematical center in Russia ever, shaped the whole of Russian mathematics; its

---

[95] In this sense Tokarenko [Tok87] cited N. B. Delone, who was, for example, the author of the first comment on Chebyshev's work about mechanism theory [Del00]. He pointed out that the break-through on this subject was reached only by the work of the later academician I. I. Artobolevski, who has also commented a lot on Chebyshev's contributions.

[96] E. g., [Gus57]—here the method of constant corrections (метод поправок), only roughly presented by Chebyshev in [Cheb54], is described.

representatives worked and taught in many of the Russian universities: in Kharkiv (Tikhomandritski, Lyapunov), Kyiv[97] (Grave), Warsaw (Sochocki, Sonin, Voronoy), Kazan (Vasilev) and Moscow (Tsvetkov).

The approximation theory did not flourish in St Petersburg, but later did through the work of Bernstein and Psheborski in Kharkiv. As we saw, there were systemic reasons for that.

Bernstein's pupil V. L. Goncharov[98] divided the (uniform) approximation theory, after methods and subjects, into three directions: an algebraic, a function-theoretical and a function-analytical one.

The function-analytical one deals with the generalization of classical results, so clearly it was missing from Chebyshev and began later, for example, with Bernstein, Akhiezer, Jackson and Haar; a first contribution to this branch might be J. W. Young's paper.[99]

The function-theoretical direction is interested in the properties of the minimal solution p_0 and the quantity

$$E_n(f) = \min_{p \subset \mathbb{P}_n} \|f - p\|.$$

Chebyshev himself only showed some inequalities it satisfies [Cheb59], but neglected questions referring to its convergence for $n \to \infty$. The latter question was decided by the Weierstraß approximation theorem [Wei85].

The pioneering investigations were those of the Markov brothers into estimates for the derivatives of polynomials.

The value of this branch of research arose with the quantitative results of Jackson and Bernstein and became the centre of the 'Constructive Function Theory' as it was called by Bernstein and by the whole Russian school.

The algebraists among the mathematicians with approximation theory investigated minimax problems with side-conditions for the coefficients of the approximating polynomials, e. g., Chebyshev's first case from „Sur les questions de minima ..." [Cheb59], that is, for fixed σ:

$$\min_{p \in \mathbb{P}_{n-1}} \max_{x \in [-1,1]} |\sigma x^n - p(x)|. \tag{3.56}$$

All of the early St Petersburg problems were posed in a similar matter.

Thus we are able to present Chebyshev's opinion in the debate about the foundations of analysis. It is clear why all his results fit into the algebraic direction.

His pupils' positions were diverse, but obviously more theoretical approaches were discussed more intensively.

---

[97] The Russian name for the Ukrainian capital is 'Kiev.'
[98] See [Gon47/2].
[99] [You07], compare section 4.4.2.

3.7.1 Decadency and Pseudo-Geometry

As often happened in history, some of the members of the St Petersburg Mathematical School had a more radical view than Chebyshev. Of course he clearly separated himself from methods and opinions of certain Western European schools, but no quotations are known where he criticized concrete contributions of specific mathematicians or even their work as a whole.

Some of his pupils however made such specific criticisms. Aleksandr Nikolaevich Korkin called the mathematical interests of Riemann and Poincaré 'decadency,'[100] and we have the following clear citation of Aleksandr Michailovich Lyapunov, which was even printed in [Lya95], where he presented his view at the St Petersburg Mathematical School, which he called the Chebyshev School. He wrote in the 1895 necrologue about the life and work of Chebyshev:

"One of his [i.e. Chebyshev's, K.G.S.] main merits as a professor was the foundation of a mathematical school, known by his name and distinguished by a special research direction. The pupils of P. L. Chebyshev continued and continue to work out the methods discovered by him. Solving the problems which they were they find new problems of the same kind. On this way step by step there arise new branches of research which will be connected forever with the name of P. L. Chebyshev. Together with this the opinions which the famous scientist was aware of in all his contributions are spreading further and further through the work of his successors.

At a time, when the admirers of the extremely abstract ideas of Riemann are burying themselves more and more in function-theoretical[101] and pseudo-geometrical investigations in spaces of four or more dimensions and go sometimes in these works so far that any possibility vanishes to recognize there any meaning for any application, not only today, but also in the future—at that time P. L. Chebyshev and his successors constantly stand on a realistic ground led by the view that only the investigations originated by applications (scientific or practical) have a value, and only theories coming from the investigation of special cases are useful.

The detailed work out of the questions which are especially important from an applied perspective and simultaneously are of special theoretical difficulty, which force the invention of new methods and their connection to the principles of science, and also the generalization of the results which we got in this way and the development of a more or

---

[100] Compare section 3.1.

[101] The word ,,функционально-теоретические" used by Lyapunov is difficult to translate, because he did not try to name a method, but wanted to disparage a mathematical direction. Therefore we tried to translate it word-for-word.

less general theory—of such kind is the direction of most of the work
of P. L. Chebyshev and the scientists who adopted his opinions."[102]

Of course it is an open question how this attitude could be harmonized
with an engagement in problems like quadratic forms which did not have
any practical applications at that time, but nevertheless were supported by
Chebyshev and deepened especially by Korkin and Zolotarev.

Why is it unseemly to ask whether a continuous function always can be
represented by a Fourier series, if it is allowed to ask for the number of primes
under a certain boundary?

Obviously common law was valid, for example Euler could not be criticized. Unfortunately it is not recorded, how Petersburg mathematicians
directly discussed these questions with their Western colleagues, especially
the 'pseudo-geometers'—the contacts had already been established.

Anyway these views contributed to the fact that for the time being St Petersburg approximation theory ended in an impasse.

---

[102] [Chebgw4, p. 19 f.:] «Главная заслуга его, как профессора, заключается
в создании той школы математиков, которая известна под его именем
и характеризуется особым направлением исследований. Ученики П. Л.
Чебышева продолжали и продолжают разработку изобретённых им
методов и при решении поставленных им задач выдвигают новые задачи
того же рода. Таким образом, малопомалу создаются новые отделы в
науке, с которыми всегда будет связанно имя П. Л. Чебышева. Вместе
с тем работами его последователей всё более и более распространяются
те взгляды, которым великий ученый оставался верен во всех своих исследованиях.

В то время, как почитатели весьма отвлеченных идей Римана всё более
и более углюбляются в функционально-теоретические исследования и
псевдо-геометрические изыскания в пространствах четырёх и большего
числа измерений, и в этих изысканиях заходят иногда так далеко, что
терается всякая возможность видеть их значение по отношению к каким-
либо приложениям не только в настоящем, но и в будущем — П. Л.
Чебышев и его последователи остаются постоянно на реальной почве,
руководясь всглядом, что только те изыскания имеют цену, которые
вызываются приложениями (научными или практическими), и только те
теории действительно полезны, которые вытекают из рассмотрения частных случаев.

Детальная разработка вопросов, особенно важных с точки зрения
приложений и в то же время представляющих особенные теоретические
трудности, требующие изобретения новых методов и восхождения к
принципам науки, затем обобщение полученных выводов и создание этим
путём более или менее общей теории - таково направление большинства
работ П. Л. Чебышева и учёных, усвоивших его взгляды.»

4

Development Outside Russia

4.1 The Mediator: Felix Klein

As we saw in his biography, P. L. Chebyshev loved to travel. His favorite destination was France, partly because he knew this language excellently. There his ideas found a big echo. The Chebyshev polynomials introduced by him in 1853 already were discussed in 1864 in Bertrand's famous calculus textbook [Bert64, §§488–491].

Nevertheless the contributions of St Petersburg mathematicians remained the only papers about uniform approximation theory for a long time. There is no contribution on this subject published earlier than 1900 and written by a mathematician who did not work or study in Russia.

One of the reasons for this might be the limited availability of the Russian papers. In 1901 H. F. Blichfeldt remarked that he had "not access to the original memoirs of Tschebycheff" [Bli01, p. 102] and Runge's scientific biographer Gottfried Richenhagen judged so after an analysis of Runge's work from the beginning of the last century [Ric85].

Felix Klein[1] knew this problem. In May 21, 1895 he gave a talk before the Göttingen Mathematical Society about Chebyshev's work on L_2-approximation. We present the protocol of this speech in full length, because it shows very clearly how Klein was also engaged in the effort to get an overview of mathematics of the time, for himself and for his colleagues.

"In a short overview the signing person talked about the work of Chebyshev according to the interpolation by polynomials.
Mainly we saw a great series of contributions, where the principle of least squares was chosen as a starting point. In 1886 Posse put together the results coming out from here in the book „Sur quelques

---

[1] Felix Christian Klein (*Düsseldorf 1849, †Göttingen 1925), student in Bonn, 1871-72 lecturer in Göttingen, 1872–1875 ord. prof. univ. Erlangen, 1875–1880 ord. prof. technical univ. München, 1880-1886 ord. prof. univ. Leipzig, 1886–1913 (emeritus) ord. prof. univ. Göttingen.

applications des fractions continues algébriques".[2] The polynomial y does not need to be ordered after increasing powers of x, but after certain partial polynomials $\omega_\nu(x)$

$$y = c_0\omega_0 + c_1\omega_1 + c_2\omega_2 + \cdots$$

resulting as approximate denominators of a simple expansion into continuous fractions. The values $c_0, c_1, c_2, \ldots$ are determined in a most simplest way so that the value of c_ν does not depend whether one wants to use other polynomials than ω_ν evaluating y. In certain special cases the general formula we got so also contains series expansions the cosinus of the variable, spherical functions and also Taylor's series, what gives us a scale for the inner meaning of these series. The expansion, which shows whether the one or the other series will be the result, depends on the fact how the weight of the observations is distributed within the interval where we have to approximate with a polynomial - constantly, like $\frac{dx}{\sqrt{1-x^2}}$, and so on. Simultaneously the ω_ν are useful for other purposes: for Newton's quadrature and for the very interesting problem of giving values for the integral $\int_b^a \Lambda(y)f(y)\,dy$ with given values of $\int_b^a f(y)\,dy, \int_b^a yf(y)\,dy, \ldots, \int_b^a y^n f(y)\,dy$.

But these expansions were not the end of the work of Chebyshev who always started from immediate practical applications and so was free from any compulsion of certain methods. Already his first work is especially remarkable. It was about the most useful dimensions of the so-called parallelogram mechanisms—mechanisms reaching an approximately linear motion of steam engines—and Chebyshev made his debut with it in the St Petersburg Savants étrangers VII in 1853. There he found the polynomial $y = a + bx + \cdots + \underline{x^n}$, which deviates according to its absolute value as little as possible from zero between ± 1. Chebyshev found that this polynomial is $\left(\frac{x+\sqrt{x^2-1}}{2}\right)^n + \left(\frac{x-\sqrt{x^2-1}}{2}\right)^n$; its largest deviation from zero is $\frac{1}{2^{n-1}}$.

There is no doubt that all these contributions of Chebyshev are especially remarkable, both under theoretical and under the aspect of an immediate numerical application. *Unfortunately until now they are hardly available and nearly unknown in Germany; we want to take all pains to make a change on this behalf.*"[3]

<u>Klein</u>.[4]

---

[2] Compare [Pos86] among the references of the present work.

[3] Italics set by us, K.G.S.

[4] [MathGö, p. 133–136]: „Der Unterzeichnende berichtete in Kurzem Überblicke über die Arbeiten von <u>Tchebyscheff</u> betr. Interpolation durch Polynome.

Since his Leipzig time Klein had been interested in Russian contributions
to different subjects in mathematics. In his Göttingen assets we find some letters which provide information about that: he corresponded with A. Markov

Wir sahen vor allen Dingen eine grosse Serie von Arbeiten, in denen das Prinzip
der Kleinsten Quadrate als Ausgangspunkt gewählt wird. Die hier entstehenden
Resultate hat Possé 1886 in dem Buche: ‚Sur quelques applications des fractions
continues algébriques' im Zusammenhang dargestellt. Das Polynom y muß nicht
nach ansteigenden Potenzen von x sondern nach gewissen Theilpolynomen $\omega_\nu(x)$
geordnet werden:

$$y = c_0\omega_0 + c_1\omega_1 + c_2\omega_2 + \ldots,$$

die sich als Näherungsnenner einer einfachen Kettenbruchentwicklung ergeben.
Es bestimmen sich dann die $c_0, c_1, c_2, \ldots$ in einfachster Weise, so zwar, dass der
Werth von c_ν gar nicht davon abhängt, ob man ausser ω_ν noch andere Theilpoly-
nome beim Auswerthen des y benutzen will. In der so entstehenden allgemeinen
Formel sind bei gew. Reihenentwickelungen nach Cosinus der Vielfachen, nach
Kugelfunktionen, auch die Taylor'sche Reihe als besondere Fälle mitenthalten,
was einen Masstab für die innere Bedeutung dieser einzelnen Reihen abgibt. Die
Entwicklung nämlich, ob die eine oder andere Reihe sich einstellt, hängt daran,
wie man sich das Gewicht der Beobachtungen im Intervall, innerhalb dessen mit
einem Polynom approximiert werden soll, vertheilt denkt, ob constant, ob als
$\frac{dx}{\sqrt{1-x^2}}$, etc. etc. Zugleich sind die ω_ν auch nach anderen Richtungen nützlich:
bei der newtonschen Quadratur und bei der sehr interessanten Aufgabe für
das Integral $\int_b^a \Lambda(y)f(y)\,dy$ bei gegebenen Werthen der $\int_b^a f(y)\,dy$, $\int_b^a yf(y)\,dy$, $\ldots$,
$\int_b^a y^n f(y)\,dy$ Grössenwerthe anzugeben.

Mit seinen Entwickelungen sind aber die Untersuchungen von T. [Chebyshev,
K.G.S.], der immer von Fragen der unmittelbaren praktischen Anwendungen aus-
geht und daher in der Wahl seiner Probleme von jedem Zwange irgend welcher
bestimmter Methoden frei ist, keineswegs erschöpft. Besonders bemerkenswert ist
gleich die erste Arbeit, mit der T. 1853 in den Petersburger Savants étrangers VII
debutierte, über die zweckmässigsten Dimensionen der sog. Parallelogrammecha-
nismen (durch welche bei den Dampfmaschinen angenäherte Geradführung erreicht wird). Es handelt sich dort um Aufsuchung des Polynoms $y = a+bx+\cdots+\underline{x^n}$,
welches zwischen ± 1 seinem absoluten Betrage nach möglichst wenig von Null
abweicht. T. findet, dass das $\left(\frac{x+\sqrt{x^2-1}}{2}\right)^n + \left(\frac{x-\sqrt{x^2-1}}{2}\right)^n$ ist; seine grösste Ab-
weichung von Null beträgt $\frac{1}{2^{n-1}}$.

Es ist kein Zweifel, dass alle diese Arbeiten von T. sowohl unter theoretischem
Gesichtspunkte als unter dem Gesichtspunkte der unmittelbaren numerischen Anwendung besonders bemerkenswerth sind. Leider aber sind dieselben bisher wenig
zugänglich und in Deutschland so gut wie unbekannt; wir werden uns alle Mühe
geben wollen, in dieser Hinsicht eine Änderung herbeizuführen."

Klein.

[MarAKl], Tikhomandritski [TiKl], Posse [PoKl] and Mlodzeevskij,[5] who spent some time in Göttingen about 1890[6] [MlKl].

A. Markov reported to him about the content of the work of Chebyshev and his pupils, in a letter from February 14, 1885 [MarAKl, Nr. 918] he referred to the application of the theory of continuous fractions to series expansions of functions and the interpolation by means of least squares and asked Klein if he was interested in applications to the theory of functions least deviating from zero, that is, to the best Chebyshev approximation.[7]

Unfortunately there is no other letter where Markov described the content of Chebyshev's work more precisely, but maybe Klein only asked Markov to send the contributions to him. The constancy of the correspondence suggests this.

In a letter of nine pages[8] from November 14, 1884 [TiKl, Nr. 24] Tikhomandritski informed Klein about the circumstances and prevailing data of Russian mathematics. He wrote which journals were published in Russia and who taught at Russian universities and who represented which subject.

Klein's rôle as a mediator[9] would soon bear fruit, as we will see. It was made easier by the edition of Chebyshev's collected works by Markov and Sonin in 1899.

---

[5] Mlodzeevskij, Boleslav Kornelievich (June 28 (July 10) 1858 (1859?)- January 18, 1923), 1892–1911 and from 1917 professor at Moscow university, 1906–1921 vice president of the Moscow Mathematical Society, 1921–1922 its president. His subjects of interest were algebraic geometry, differential geometry and the theory of functions of a real variable (about this he lectured since 1900!).

[6] This we took from his letter to Felix Klein from October 28 (November 9), 1891, where he gratefully remembered his time in Göttingen.

[7] [MarAKl, Nr. 918, S. 2/3]: ,,Il me parait, que l'application des fractions continues au calcul approché des intégrales est suffisament connue.

Peut être, leurs applications au développement des fonctions en séries et à l'interpolation par la méthode des moindres carrés sont moins connues; ce sont précisement quelques formules de M. Tchébychef. [...]

Vous avez peut-être en vue d'autres fractions continues et rélativement à cela d'autres questions, par example: des fonctions qui s'écartent le moins possible de zéro, ou bien de l'intégration sous forme finie. J'aurais pu parler sur la première chose."

[8] The entry Cod. Ms. F. Klein 12, Nr. 24 also contains a supplement about Polish, Czech and Hungarian journals, which obviously (by handwriting) was collected by others.

[9] We want to remark that Klein's engagement was ignored in the Soviet Union. The Soviet encyclopedia wrote about him: "K[lein] described it [the history of mathematics] tendentiously exaggerating the merits of German mathematicians. In particular, he concealed the work of P. L. Chebyshev." ([VaVv49-57, vol. 21, p. 403]: «К. излагал её тенденциозно, преувеличивая заслуги немецких математиков. В частности, К. замалчивал работы П. Л. Чебышева.»

4.2 Blichfeldt's Note

The first work appearing beyond Russia and dealing with approximation theory in the sense of Chebyshev was Blichfeldt's short note [Bli01] published in 1901 in the USA. Blichfeldt[10] noted that he obviously found a proof of the alternation theorem:

> "Let us by the "maxima" of $[f(x)]^2$ in the interval $a \leq x \leq b$ understand those only which are equal to L^2, the greatest value of $[f(x)]^2$ in the given interval. If we classify these maxima as *positive* or *negative* according as the corresponding value of *f(x)* is $+L$ or $-L$, and plot the curve *y=f(x)*, we shall find at least n alternations of the two kinds of maxima in the given interval."

Here Blichfeldt did not prove the theorem, he only mentioned that a construction is possible presenting a shift argument. It might be that Blichfeldt did not give a complete proof only because he did not know all of Chebyshev's work.[11]

We will see that there were difficulties in the first attempts of a proof of the alternation theorem, therefore we are not able to name him as the first who proved the alternation theorem, since we do not know his proof.

4.3 Kirchberger's Thesis

We have already reported that Felix Klein tried to make the work of Chebyshev and his pupils public in Göttingen.

So it seems to be a consequence of this effort that just there Chebyshev's ideas got a new foundation—immediately after the 1899 edition of Chebyshev's collected work by Andrej Andreevich Markov and Nikolay Jakovlevich Sonin Paul Kirchberger[12] finished the thesis 'About Chebyshevian Approximation Methods' („Ueber Tchebyschefsche Annäherungsmethoden" [Kir02] which he defended in 1902 under supervision of David Hilbert.

---

[10] Hans Frederick Blichfeldt (*Illar (Denmark) 1873, †Palo Alto (USA) 1945). His mathematical main interests were group theory and number theory.

[11] [Bli01, Footnote on p. 102]: "The writer has not access to the original memoirs of Tschebycheff, in which this property may have been indicated."

[12] There is not much known about Paul Kirchberger (1878-?). In his curriculum vitae as part of his thesis he pointed out the following data: Born June 23, 1878 in Niederlahnstein (Germany), grammar school in Weilburg, 1897-1900 studies in Berlin, 1900-1902 Studies in Göttingen, 1903 assistant teacher in Fulda (this information was taken from [Beri03/04]), 1907 senior teacher (Oberlehrer) in Charlottenburg (Berlin) [Beri07]. In Berlin he stayed at least until 1922, as comes out from a letter he addressed to D. Hilbert on the occasion of his 60th birthday [KiHi].

Before we analyse this work more precisely, we want to have a look i nto
the report the supervisor David Hilbert[13] wrote about the thesis:

"After the definition of the Chebyshev approximating function for a
finite interval, the proof for the existence of this function is presented,
to be precise, simultaneously for the case of several variables, what
Chebyshev had not done before. Then mainly Chebyshev's results
according to the properties of the approximating function are presented in a simplified form (chapter I). Now the candidate considers
the Chebyshev approximating function for a given analytic function in
its dependence on the interval length $2h$, states its analytical character
according to h and finally gets to a convergent algorithm to calculate
the approximating function by means of a string of divisions (chapter II). After expanding Chebyshev's problem in chapter III adding
some remarkable initial and side-conditions, in chapter IV the candidate passes to formulate and prove some auxiliary theorems about
convex polyhedra. These theorems are partly equal to some general
theorems Minkowski found quite some time ago but did not publish.
They serve the candidate to develop the theorems of chapter V, which
represent the main result of the investigations of the author and by
which a complete expansion and generalization of Chebyshev's theory
is reached to the case of approximating functions of several variables.
The work, especially the results of chapter V are of a large scientific
value; the presentation is careful and skillful."[14]

---

[13] David Hilbert (*1862 Königsberg, †1943 Göttingen), 1880–1885 studies in
Königsberg, 1886-1892 Privatdozent univ. Königsberg, 1892–1895 extraord. prof.
univ. Königsberg, 1895–1930 (emeritus) ord. prof. U Göttingen.

[14] Here we cite the complete report as it can be found in the doctorate's al-
bum ('Promotionsalbum') of the philosophical faculty (Archive of Göttingen uni-
versity, Phil. Fak., Nr. 188b, Nr. 11): „Nach Definition der Tchebyschef'schen
Näherungsfunktion bezüglich eines endlichen Intervalles wird der Beweis für die
Existenz dieser Funktion erbracht, und zwar zugleich für den Fall mehrerer
Variablen, was von Tchebyschef noch nicht geschehen war. Sodann werden im
Wesentlichen die Tchebyschef'schen Resultate betreffend die Eigenschaften der
Annäherungsfunktion in vereinfachter Form dargestellt (Kap. I). Nunmehr betrachtet der Candidat die Tchebyschef'sche Näherungsfunktion für eine gegebene
analytische Funktion in ihrer Abhängigkeit von der Intervallänge $2h$, stellt den
analytischen Charakter derselben in Bezug auf h fest und gelangt schliesslich
zu einem convergenten Verfahren zur Aufstellung der Annäherungsfunktion mit-
telst einer Kette von Divisionen (Kap. II). Nachdem in Kap. III die Tchebyschef'sche Fragestellung noch durch Hinzunahme gewisser merkwürdiger Anfangs-
und Nebenbedingungen erweitert wird, geht der Candidat in Kap. IV dazu über,
gewisse Hülfssätze über convexe Polyeder aufzustellen und zu beweisen. Diese
Sätze decken sich zum Teil mit gewissen allgemeinen Theoremen, die Minkowski
schon seit längerer Zeit gefunden, aber noch nicht publicirt hat. Sie dienen dem
Candidaten zur Entwickelung der Sätze in Kap. V, die das Hauptergebnis der

Of course Kirchberger was allowed to sit at the examination after such a report. July 17, 1902 he passed it. The examiner of mathematics was Hilbert, another member of the examination committee was Klein.[15]

Although Hilbert emphasized the special meaning of the last two chapters—their results would later be published in the 'Mathematische Annalen' [Kir03]—we want to restrict ourselves to the first three chapters because they immediately continue Chebyshev's work.

4.3.1 Existence, Uniqueness and Continuity

As we have already shown several times, existence and uniqueness theorems are not important for the protagonists of the St Petersburg Mathematical School— Chebyshev clearly said that he develops a theory to solve concrete problems. So questions of existence and uniqueness are automatically answered because his solutions are concrete polynomials of a given degree.

On the other hand Kirchberger was influenced by the Berlin and Göttingen schools—he studied both in Berlin and in Göttingen. In his curriculum vitae in the appendix of [Kir02] he emphasized Frobenius' influence on him and besides others he also mentioned Fuchs und H. A. Schwarz. So it is no wonder that he tried to build secure foundations for the theory of best approximation.

4.3.1.1 Return to Weierstraß' Fundamental Theorem

To prove the existence of a best approximation Kirchberger stated that it suffices to show that all coefficients of the minimal sequence are uniformly bounded, that is if $f \in C[a, b]$ is the function to be approximated,

$$\left(p^{(m)} \right)_{m \in \mathbb{N}} \subset \mathbb{P}_n$$

the minimal sequence and for all $m \in \mathbb{N}$:

$$p^{(m)}(x) = \sum_{i=0}^{n} a_i^{(m)} x^i, \quad x \in [a, b],$$

there exists a bound $M \in \mathbb{R}$ so that

$$\sup_{m,i} |a_i^{(m)}| \leq M.$$

Untersuchungen des Verfassers darstellen und durch die eine vollkommene Ausdehnung und Verallgemeinerung der Tchebyschef'schen Theorie auf den Fall von Annäherungsfunktionen mehrerer Veränderlicher erreicht wird.

Die Arbeit besonders die Ergebnisse von Kap. V sind von grossem wissenschaftlichen Wert; die Darstellung ist sorgfältig und geschickt."

[15] Ibid.

Kirchberger argued correctly that with this it is clear that the minimal deviation

$$L(a_0, \ldots, a_n) = \inf_{x \in [a,b]} |f(x) - \sum_{i=0}^{n} a_i x^i|$$

is a function of $n+1$ variables from the $n+1$-dimensional cube $[-M, M]^{n+1}$, and therefore reaches its minimum there according to a theorem of Weierstraß.

4.3.1.2 Proof of the Alternation Theorem

In the following section Kirchberger refined Chebyshev's theorem (see Theorem 2.1 from page 38).

We again consider the general approximation problem for $n+1$ parametres $p_0, \ldots, p_n$, with the error function $F(., p_0, \ldots, p_n) \in C[-1, 1]$. Let $x_1, \ldots, x_\mu \in [-1, 1]$ be the deviation points of F.

Then there holds Chebyshev's following theorem formulated by Kirchberger:

Theorem 4.1 *If the deviation* $\|F(., p_0, \ldots, p_n)\|$ *is minimal, then the system of equations*

$$\lambda_0 \frac{\partial F}{\partial p_0}(x_1) + \lambda_1 \frac{\partial F}{\partial p_1}(x_1) + \ldots + \lambda_n \frac{\partial F}{\partial p_n}(x_1) = s_1$$
$$\lambda_0 \frac{\partial F}{\partial p_0}(x_2) + \lambda_1 \frac{\partial F}{\partial p_1}(x_2) + \ldots + \lambda_n \frac{\partial F}{\partial p_n}(x_2) = s_2 \qquad (4.1)$$
$$\vdots \qquad\qquad\qquad\qquad \vdots \quad \vdots$$
$$\lambda_0 \frac{\partial F}{\partial p_0}(x_\mu) + \lambda_1 \frac{\partial F}{\partial p_1}(x_\mu) + \ldots + \lambda_n \frac{\partial F}{\partial p_n}(x_\mu) = s_\mu$$

with $\operatorname{sign} s_i = \operatorname{sign} F(x_i)$ *for all* $i = 1, \ldots, \mu$ *is not solvable.*

This theorem slightly differs from Chebyshev's theorem because it takes signs into account. The proof, however, does not become more difficult.

The theorem immediately implies in the case of polynomial approximation that there must be at least $n + 2$ deviation points if the error function is minimal, and so we get Chebyshev's result. Then the system of equations has the shape

$$\lambda_0 + \lambda_1 x_1 + \ldots + \lambda_n x_1^n = s_1$$
$$\lambda_0 + \lambda_1 x_2 + \ldots + \lambda_n x_2^n = s_2 \qquad (4.2)$$
$$\vdots \qquad\qquad \vdots \quad \vdots$$
$$\lambda_0 + \lambda_1 x_\mu + \ldots + \lambda_n x_\mu^n = s_\mu.$$

If we now assume that $H := \sum_{i=0}^{n} \lambda_i \iota^i$ is a polynomial with zeros between x_{s_j} and $x_{s_{j+1}}$, whenever there is a change in sign between s_j and s_{j+1}, then we see that it is possible to construct such a polynomial if we have not more than n changes of signs of the error function. So Chebyshev's theorem in the formulation of Kirchberger implies the existence of $n + 2$ *alternating* deviation points.

But the approach via the system of equations has the disadvantage that we firstly have to assume the finiteness of the number of deviation points. As Chebyshev already did, so also Kirchberger argued with the (already famous) shift, namely that, if the systems of equations (4.1) or (4.2) are solvable, a function ρ from the space of approximating functions can be taken with the property

$$\|F(., p_1, \ldots, p_n) - \rho\| < \|F(., p_1, \ldots, p_n)\|.$$

The diminishing of the error is possible in small neighborhoods of the deviation points. Beyond them the value of the difference $|F(x, p_1, \ldots, p_n) - \rho(x)|$ may become larger, but will not exceed the maximum error, if x is not a deviation point not having been taken into account. In the case of the finiteness of deviation points this won't be a problem because then all deviation points are part of the system of equations.

Kirchberger pointed out about this problem that, if we have whole deviation intervals, the respective neighborhoods can be suitably expanded, which is absolutely correct. But the case of an infinite number of deviations not covering whole intervals is not sufficiently treated with this remark.

We have such a case when approximating $f : [0, 1] \to \mathbb{R}$ with

$$f(x) = \begin{cases} -\frac{1}{2} + \frac{1}{\sin(1)} \left| x \sin(x^{-1}) \right| & : \quad x \neq 0 \\ -\frac{1}{2} & : \quad x = 0 \end{cases} \tag{4.3}$$

by constant functions.[16] Its best approximation is the zero function, and the only positive deviation point will be $x = 1$, negative deviation points are all points $x_k = 1/(k\pi)$, $k \in \mathbb{N}$. Figure 4.1 outlines the oscillating behaviour of the error function f.

The fact that now the signs of the deviation points had also been taken into account in Theorem 4.1 made possible a complete characterization of the best approximating polynomials because the alternation of the deviation points of the error function is also sufficient for the minimality. Kirchberger was the first to prove this fact, presenting the (now classical) contradictory proof:

We assume that $f \in C[a, b]$ is to be approximated, and let $p \in \mathbb{P}_n$ be the polynomial so that $f - p$ has at least $n + 2$ alternating deviation points $x_1, \ldots, x_{n+2}$, and let $q \in \mathbb{P}_n$ be a polynomial with

---

[16] For more detailed considerations compare [Stef94]. A respective proof has to avoid the system of equations and has to be made with topological arguments.

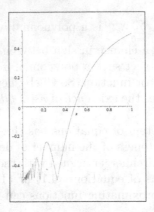

Figure 4.1. *An error function with an infinite number of deviation points not containing a whole interval.*

$$\|f - q\| < \|f - p\|.$$

Then we have in the deviation points:

$$q(x) - p(x) > 0, \quad \text{if} f(x) - p(x) > 0, \quad \text{and}$$
$$q(x) - p(x) < 0, \quad \text{if} f(x) - p(x) < 0.$$

But because of the alternating order of the deviation points $q-p$, a polynomial of degree n would have $n + 1$ zeros.

And so the question of uniqueness was answered. Also here Kirchberger was the first to prove it—with the mentioned gap.

4.3.2 Continuity of the Operator of Best Approximation

The fifth paragraph of the first chapter was devoted to the question if the error of approximation is changing continuously, or, if the operator

$$E_n : C[a, b] \to \mathbb{R}$$
$$f \mapsto E_n(f)$$

is continuous in f.

Kirchberger proved this fact with the alternation theorem strictly speaking, only for almost all continuous functions—as we saw, there exist functions for which Kirchberger's theorem does not give an answer.

Let f be a function with its best approximation ϕ and let f_ε be another function with best approximation ϕ_ε, for which there holds

$$\|f - f_\varepsilon\| < \varepsilon,$$

so it is easy to see that within a small neighborhood $U(x)$ of a deviation point x the difference $\|\phi - \phi_\varepsilon\|_{U(x)}$ also will become small because otherwise the polynomial of degree n, $\phi - \phi_\varepsilon$ would have n extrema.

Now we have a partial interval $U(x) \subset [a, b]$, where the polynomial $\phi - \phi_\varepsilon$ becomes arbitrarily small. Thus, its coefficients also become arbitrarily small, and with them the whole polynomial on the whole interval.

4.3.3 Rational Approximation

Kirchberger's investigations also cover the approximation of continuous functions by rational functions with a given sum of the degrees of numerator and denominator (Chebyshev's third case), but here his considerations did not convince as much as they did in the linear case.

Trying to prove the existence of the best rational approximation, he made a mistake.

He argued correctly that the denominator of a completely cancelled best approximating rational function cannot have zeros in the interval $[a, b]$, but he did not see that just the rational functions having a cancelled representation (the non-normal ones) are points of discontinuity of the operator of best approximation.[17] But he explicitly used the continuity of the operator.[18]

---

[17] see, e. g. the theorem of Cheney and Loeb in [ChLo66].

[18] In his considerations Kirchberger restricted the parametric space and considered only rational functions

$$\frac{\sum_{i=0}^{m} q_i x^i}{\sum_{i=0}^{n} p_i x^i},$$

whose coefficients lie in a bounded domain. He designated by L the operator of best approximation.

Now he stated "In this domain L can be considered as a continuous function of p and q, because the only points of infinity are the points of infinity which are not of importance for our minimal problem. So, L takes its minimum by Weierstraß' theorem. If for that minimum there holds $p_0 \neq 0$, the theorem has been proved. If, however, $p_0 = 0$, it also must be $q_0 = 0$, because otherwise for $x = 0$ the approximating function, and so L would become infinitely large and a minimum could not take place. So we can now divide numerator and denominator by x and so diminish the degree of both by 1. So we pass to a space whose dimension is lower by 2 and consider the analogous cube. Thus, we can continue [...]." ([Kir02, p. 22:] „L kann in diesem Gebiet als stetige Funktion der p und q betrachtet werden, denn die einzigen Unendlichkeitsstellen sind die Unendlichkeitsstellen, die für unsere Minimumsfrage nicht in Betracht kommen. L nimmt also sein Minimum nach dem Weierstraß'schen Satz an. Ist für dieses $p_0 \neq 0$, so ist der Satz bewiesen, ist aber $p_0 = 0$, so muss notwendig auch $q_0 = 0$, weil sonst für $x = 0$ die Annäherungsfunktion, also auch L unendlich würde, ein Minimum also sicher nicht stattfände. Wir können alsdann Zähler und Nenner durch x teilen und so den Grad beider um 1 erniedrigen. Wir gehen deshalb zu einem um

Nevertheless the proposition was right, as Walsh showed in 1931 [Wal31].

For the case of a normal (non-cancelled) rational approximation Kirchberger then proved the alternation condition.

4.3.4 A Discrete Approximation Problem

At the end of the first chapter Kirchberger presented an interesting application.

Given six points on the plain he tried to find the conic section being next to them.

He then showed a necessary and sufficient condition for this situation, namely that the distances of the six points to the conic section must be equal and that the points have to lie alternating left and right from it, where we have to understand 'alternating' suitably (e. g., following the arc).

We see that this is a new interpretation of Laplace's problem from section 1.2.

Kirchberger correctly mentioned that only special two-dimensional problems can be treated in this way, because in general we cannot assume uniqueness in higher dimensions.[19]

In a further section he calculated the best approximating polynomial for a table of values. Of course he was not able to give an algorithm for the general problem and closed the chapter with the words

> "The interpolation problem gives the main results of Chebyshev's approximation methods. But the problem is reduced here to the solution of linear equations."[20]

4.3.5 An Algorithm for the Approximate Determination of the Best Approximation of Real-Analytic Functions

Kirchberger now carefully showed how it is possible to use the Chebyshev polynomials for the determination of best approximation of analytic functions. Presumably Chebyshev himself had solved this problem, as his explanations from „Théorie des mécanismes..."[Cheb54] assume, but we do not find such an algorithm among his published documents.

---

2 Dimensionen niedrigeren Raum und betrachten in diesem den analogen Würfel. Auf diese Weise fahren wir fort [...]")

We clearly see how he followed a wrong track, since the minimum can be reached beyond a given cube because the cancellation of rational functions is a discontinuous operation.

[19] Compare the theorems of Haar [Haa17] and Mairhuber [Mai56].

[20] [Kir02, p. 30:]„Das Interpolationsproblem liefert also die wesentlichen Resultate der Tchebyschefschen Annäherungsmethoden. Das Problem reduziert sich aber hier auf die Auflösung linearer Gleichungen."

We write (as Chebyshev already did, compare section 2.3.2) the given analytic function $\phi \in C[-h, h]$ depending on the boundaries of the interval

$$\phi(z) = \sum_{i=0}^{\infty} k_i (hz)^i$$

$$= \sum_{i=0}^{n} k_i h^i z^i + h^{n+1} \sum_{i=n+1}^{\infty} k_i h^{i-n-1} z^i$$

$$=: p(z) + h^{n+1} \psi(z).$$

We know,[21] that the best approximation for ϕ then has the form

$$p + h^{n+1} V,$$

where $V := \sum_{i=0}^{n} p_i z^i$ is the best approximation for ψ.

Using the characteristic equations in the deviation points $z_1, \ldots, z_{n+2}$ of $V - \psi$,

$$V(z_2) - \psi(z_2) = L$$

$$\vdots$$

$$V(z_{n+1}) - \psi(z_{n+1}) = (-1)^{n+1} L$$
$$V'(z_2) - \psi'(z_2) = 0$$

$$\vdots$$

$$V'(z_{n+1}) - \psi'(z_{n+1}) = 0$$
$$z_1 = -1$$
$$z_{n+2} = 1,$$

Kirchberger saw that their solutions $p_0, \ldots, p_n, L, z_2, \ldots, z_{n+1}$ could be expanded by powers of h.

Now he determined successive approximations for V, increasing the order of h step by step, as Chebyshev had already suggested. The closeness of both texts let us assume that Chebyshev indeed had already found the algorithm. But he obviously preferred to give closed representations of best approximations, as his solutions and the solutions of his pupils let us presume. Therefore it might be that Chebyshev did not think that the algorithm was of high importance.

4.3.6 Approximation under Side-Conditions

Kirchberger finished his investigations into the best approximation of continuous functions by discussing two special problems; firstly he considered the

---

[21] See equation (32) and the following remarks.

case of one interpolatory side-condition, secondly he dealt with the determination of the monotone polynomial of degree n with given first coefficient which least deviates from zero. This problem we have already carefully investigated, and since Kirchberger did not add any new results except for an existence theorem, we don't want to discuss it here.

The interpolatory side-condition was the following one: The polynomial of best approximation to f should also interpolate f at the endpoints of the interval.

Kirchberger proved an alternation theorem for this case: It is necessary and sufficient for a solution of this problem that the error function has n alternating deviation points (not $n + 2$ as in the general problem), because of the two additional equations the number of deviation points was diminished by two, as could be expected.

4.3.7 Foundations for Chebyshev's Methods

Although David Hilbert did not especially praise Kirchberger's above-mentioned results, according to the one-dimensional approximation of continuous functions they also were of special use. These results added analytic considerations to St Petersburg's algebraic results, and firstly Kirchberger proved theorems about the existence, uniqueness and continuity of the occurring solutions.

St Petersburg purists would have denied that Kirchberger's results gave new insights. But among them there was an alternation theorem by means of which he could solve a new discrete approximation problem and an algorithm to determine the best approximation of an analytic function. As we have already mentioned, presumably Chebyshev had developed a similar one before, but had never published it in such an explicit form. Kirchberger's discrete approximation problem had no forerunner in St Petersburg, but could be easily solved by the characteristic equations Chebyshev had already discussed in 1853. The alternation theorem itself could also easily be derived from Chebyshev's considerations—Kirchberger himself did it in such way.

But in spite of all this scepticism we should make clear again that the mathematical theory of the best approximation remained an incomplete structure in St Petersburg because they did not sufficiently pay attention to the approximation of continuous functions and only were concentrated in special problems. Therefore Kirchberger's work was indeed a milestone in the development of the theory and would later be recognized by Borel.[22]

---

[22] In his Leçons [Bor05] Borel wrote about Chebyshev's approximation method: [Bor05, Footnote on p. 82]: "La méthode de Tchebicheff a été reprise et rendue rigoureuse par M. Paul KIRCHERBERGER [!], *Inaugural-Dissertation: Ueber Tchebychefsche Annäherungsmethoden,* Göttingen, 1902. Nous avons utilisé dans ce qui suit cet important travail."

4.4 Other Non-Quantitative Contributions

Some of the gaps Kirchberger left would be closed soon.

4.4.1 Borel

Topology entered approximation theory via Émile Borel[23] in his 1905 „Leçons sur les Fonctions de Variables Réelles et les Développements en séries de Polynomes."

This had a special advantage for the alternation theorem. In its proof Borel could free himself from the deviation points of the error function and so from Chebyshev's system of equations (2.31) and thus avoid the trap Kirchberger ran into.

Borel divided the interval $[a, b]$ into extremal segments containing at least one deviation point and into parts where the error function remains clearly smaller than the maximum error.

These extremal segments now contain all points in a neighborhood of a deviation point where the error function has values which are larger than the boundary for the other non-extremal intervals.

More exactly: Let L be the maximum error for a given error function $f - p$ in $[a, b]$ and $x_1, \ldots, x_k$ be a k-alternant, then we define with $0 < \varepsilon < L$ the following subdivision for $[a, b]$:

$$x \in A \iff |f(x) - p(x)| \leq \varepsilon$$
$$x \in B \iff |f(x) - p(x)| \geq \varepsilon.$$

Because of the alternation property there are now k disjoint partial intervals in B. If now $k < n + 2$, then it is possible to determine a polynomial of degree n, whose zeros lie between these partial intervals. By means of this polynomial we can now diminish the maximum error as Chebyshev and Kirchberger did.

For the construction of a function with smaller norm it is important that the sets A and B are closed. There topology played its decisive rôle. Borel discussed this fact carefully.

Because of the skilful construction of the extremal segments, Borel avoided all pathological cases and did not have to find special arguments for special cases, as Kirchberger had to do.

But also Borel made a small, but fundamental, mistake in the final estimate for the value of the new error function. He correctly showed that in the extremal segments B the error function can be diminished, let's say by δ. Then he stated that also in the remaining intervals A the error function must remain smaller than $L - \delta$, but this is wrong in general.

---

[23] Félix Édouard Justin Émile Borel (*Saint Affrique (France) 1871, †Paris 1956), 1896–1909 prof. at the École Normale Supérieure in Paris, 1909–1924 prof. at the Sorbonne in Paris, 1924–1940 naval minister of the French Republic.

Presumably this was a careless mistake, but nevertheless there remained a gap, since of course it could be that his special construction caused jumps beyond the extremal segments B.

Two years later John Wesley Young showed that this was not the fact: the mistake could easily be repaired.

4.4.2 Young's Systems

In 1907 Young[24] published the contribution "General Theory of Approximation by Functions Involving a Given Number of Arbitrary Parameters" [You07]. This work was the first work after Chebyshev's "Questions about Minima..." [Cheb59] which abstracted from the uniform approximation by special families of functions like algebraic, trigonometric polynomials or rational functions.

Young was the first to define linear systems of continuous functions whose most important property they have in common with polynomials is: *Every non-trivial function from such an n-dimensional function space has not more than n − 1 zeros.*

Today we call such systems *Haar function spaces* because only those systems have the property that any continuous function can be approximated uniquely in such a function space, as Haar[25] could show in 1917 [Haa17]. Often they are also called *Chebyshev function spaces.*

Young showed the existence and uniqueness of the best approximation by elements of such a function space. The uniqueness was shown again via an alternation theorem.

He used there Borel's arguments and was able to avoid his mistake. But unfortunately also Young made a mistake. He implicitly stated that there exists a constant in every one of the function spaces defined by him, but it is only evident that they contain a strictly positive function (which by the way is sufficient to use the classical shift argument and to prove the alternation theorem).

But nevertheless at last we got a valid proof of the alternation theorem for function spaces containing constant functions like the polynomials.

---

[24] John Wesley Young (*Columbus/Ohio (USA) 1879, †Hanover, New Hampshire (USA) 1932), studied at the Ohio State University, 1903–1905 prof. Northwestern univ., 1905–1908 prof. Princeton univ., 1908–1910 prof. Illinois univ., 1910–1911 prof. Kansas univ., 1911–1932 prof. Dartmouth College, 1928–1930 vice president of the AMS, mainly worked in geometry.

[25] Alfréd Haar (*Budapest 1885, †Szeged 1933), 1904–1909 studies in Göttingen, 1909–1912 lecturer in Göttingen, 1912–1917 extraordinary prof. univ. Kolozsvár (Austria-Hungary, today Cluj-Napoca/Romania), 1917 ord. prof. univ. Kolozsvár, 1918–1933 ord. prof. U Szeged.

4.4.3 Trigonometric Approximation

Simultaneously with the publication of Young's text a work of Fréchet[26] appeared where the question of approximation of continuous functions by trigonometric polynomials was discussed [Fre07]. Here he transferred some of Kirchberger's results to this case, e. g., an existence theorem and a theorem about the continuity of the operator of trigonometric approximation. Additionally he pointed out that the coefficients of the sequence of best approximations uniformly converge to the coefficients of the Fourier series of the function to be approximated.

After another year he generalized these results, taking into account Borel's results, and extended them with a uniqueness and an alternation theorem. As an example he carefully discussed again the trigonometric approximation [Fre08].

Probably his results were discovered independently from Young's observations, but Fréchet was also not able to do without an interpolatory side-condition similar to that of Young (and Haar) to prove uniqueness and alternation theorems.

Leonida Tonelli[27] added an extensive book to these contributions in the same year [Ton08]. There he carefully collected the known results and added some new theorems about the approximation of functions of two variables.

Another work written about Chebyshev's approximation problem for functions of two variables is Sibirani's work from 1909 [Sib09].

4.5 On Convergence and Series Expansions

Weierstraß' approximation theorem of 1885 had not been mentioned by St Petersburg contributors to approximation theory, especially because of their disapproval of a deep discussion of the foundations of analysis.

So all the new impulses for approximation theory this fundamental theorem gave were developed beyond Russia.

4.5.1 Weierstraß' Approximation Theorem

Of course not only Russians were interested in questions for a good approximation of complicated functions, for example those which cannot be represented in a closed form.

---

[26] Maurice René Fréchet (*Maligny (France) 1878, †Paris 1973), 1910–1919 ord. prof. univ. Poitiers, 1920–1927 ord. Prof. univ. Strasbourg, 1928–1948 ord. prof. Sorbonne.

[27] Leonida Tonelli (*Gallipolli (Italy) 1885, †Pisa 1946), 1913 prof. univ. Cagliari, 1914–1922 univ. Parma, 1922–1930 univ. Bologna, 1930–1939 univ. Pisa, 1939–1942 univ. Rome, 1942–1946 univ. Pisa, worked on variational calculus and Fourier Analysis.

The interpolation formulae by Newton and Lagrange were well-known for a long time and simple representations like Taylor's formula and the Fourier series were often used.

Now Weierstraß' approximation theorem gave the hope that it would be possible to show that some of these algorithms converge, that they were able to approximate continuous functions arbitrarily, exactly because Weierstraß could show that both any continuous function defined on a closed interval could be arbitrarily well approximated by algebraic polynomials and any continuous 2π-periodical function could be arbitrarily well approximated by trigonometric polynomials.

It was known that Taylor's formula and the Fourier series could not give convergent sequences and were not useful in this sense. Cauchy's 1823 counterexample[28] for the Taylor series was the function

$$f(x) := \begin{cases} e^{-\frac{1}{x^2}}, & x \neq 0 \\ 0, & x = 0 \end{cases}$$

In 1867 Riemann gave an example of an integrable function with a divergent Fourier series.[29] It is the function

$$f(x) := \frac{d}{dx}\left(x^t \cos \frac{1}{x}\right) \quad (0 < t < \frac{1}{2}, \quad x \in (0, 2]).$$

After another nine years Du Bois-Reymond added a very complicated example of a continuous function with a divergent Fourier series [DuB76].

Unfortunately also the hope to be able to show the general convergence of Lagrange's interpolation formula was disappointed, as Méray [Mér96] and Runge [Run01] showed that in the case of equidistant knots it is possible to find continuous functions for which Lagrange's algorithm does not uniformly converge. Runge even found a condition which shows that even not every real-analytic function guarantees the convergence of this interpolation procedure.[30]

In 1914 Faber [Fab14] even aggravated this result showing that for every choice of knots $\{x_1^n, \ldots x_n^n\}, n \in \mathbb{N}$ there can be constructed a function for which Lagrange's algorithm diverges.

So the aim of Western European mathematicians was at first to get more handy versions of the proof of Weierstraß' for a better understanding of it and to get suitable, convergent sequences of algebraic or trigonometric polynomials.

Weierstraß'[31] fundamental idea (from [Wei85]) was the representation of a continuous function f in the form

---

[28] Cited after [Vol87, p. 209].

[29] Cited after [Vol87, S. 211].

[30] Runge's convergence criterion (from [Run01]) explicitly demanded the analyticity in a special complex domain. Compare with this the discussion in [Ric85].

[31] Karl Theodor Wilhelm Weierstraß (*Ostenfelde (Germany) 1815, †Berlin 1897), studies in Bonn ('Kameralistik'—a Prussian state accounting standard)

$$f(x) = \lim_{k \to 0} \frac{1}{k\sqrt{\pi}} \int_{-\infty}^{\infty} f(t) e^{-\left(\frac{t-x}{k}\right)^2} dt. \tag{4.4}$$

Other authors would later use such ideas of representing an arbitrary function by such an integral expression, but usually in a simpler form.

The first elementary proofs allowing a simple and applicable construction of sequences of approximating functions were those of Lipót Fejér for the trigonometric and of Edmund Landau [Lan08] for the algebraic case.[32] Fejér's work will be discussed in the next section because of its connection with Fourier series.

Probably the simplest proof of Weierstraß' theorem was given by Sergey Natanovich Bernstein [Bern12/1], which we also will comment on later.

4.6 Fejér and Runge

In their work about approximation theory, Chebyshev and his pupils concentrated on the computation of closed formulae for solutions of extremal problems. Their further aim[33] was to find a formula which allows us to determine the best approximating algebraic polynomial of given degree for any real-analytic function. Besides the general problem they discussed problems adding some side-conditions to the coefficients of the regarded polynomials.

Except for the complicated algorithm already mentioned by Chebyshev and later carefully discussed by Kirchberger, which determines solutions approximatively, there could be given only solutions for two given coefficients.

Furthermore we state that questions about convergence had hardly played an important rôle in their considerations. Maybe the clearest example is Chebyshev's remark[34] about Lagrange's interpolation algorithm, from which Runge's theorem trivially follows, that for any real-analytic function the sequence of Lagrange interpolants converges, if the zeros of Chebyshev polynomials are taken as the knots of the sequence.

Chebyshev's contributions to representation theory may be pioneering because of the discovery of orthogonality as the most important property of approximating algebraic polynomials in $L_2(\rho)$, but neither he nor any of his pupils ever investigated the question whether an arbitrary continuous function

and Münster (mathematics), 1840 state examination, 1842–1848 teacher in Deutsch-Krone (today Wałcz/Poland), 1848–1855 teacher in Braunsberg (today Braniewo/Poland), 1854 doctorate at Königsberg univ., 1856 prof. commercial institute Berlin, 1857–1864 extraord. prof. univ. Berlin, 1864-1890 ord. prof. Berlin.

[32] For a careful discussion of the early proofs of Weierstraß' approximation theorem and related questions compare the beautiful work of Allan Pinkus [Pin99]. Landau's work is discussed in the very careful and extensive overview by Butzer and Stark [BuSt].

[33] We remember again the formulations from [Cheb57] and [MarA06].

[34] Compare the discussion in section 2.4.6 on page 47.

can be represented by a series whose members come from these orthogonal polynomial spaces. This was the determining question of the discussion about the foundations of mathematics originated by the problem of the solutions of the wave equation, the 'swinging-string problem.'

Here the circle is closed, for, as we know, questions about mathematical concepts were not important for Chebyshev and the St Petersburg Mathematical School.

So the links between these different opinions about mathematics would be made in the last century, as, for example, did Fejér.

4.6.1 Summable Fourier Series

The pioneering result for the representation of continuous functions by Fourier series was given by Fejér[35] [Fej00].

The question of the general possibility to represent a continuous function by its Fourier series was negatively answered by the counterexample constructed by Bois-Reymond [DuB76] and a one-century-long discussion had ended.[36]

Therefore Fejér's following result was a sensation and theoretically enormously satisfactory: *The Fourier series of any continuous 2π-periodical function is summable.* So for the first time it was possible to relate to any function a sequence of trigonometric polynomials easy to compute and converging to this function. Weierstraß' theorem got a proof allowing a practical exploitation:

Definition 4.2 *Let $f \in L_2[-\pi, \pi]$. With $S_n(f, x)$ we want to name the nth partial sum of its Fourier series in $x \in [-\pi, \pi]$, so there holds for all $n \in \mathbb{N}$:*

$$S_n(f, x) = \frac{a_0}{2} + \sum_{k=1}^{n} a_k \cos kx + b_k \sin kx$$

with the usual coefficients

$$a_k := \frac{1}{\pi} \int_{-\pi}^{\pi} f(x) \cos x \, dx \qquad b_k := \frac{1}{\pi} \int_{-\pi}^{\pi} f(x) \sin x \, dx,$$

$k = 1, \ldots, n$. *The mean values*

---

[35] Lipót Fejér (until 1900 Leopold Weiss) (*Pécs 1880, †1959 Budapest), 1897–1902 studies in Budapest and Berlin, 1901–1905 repetitor univ. Budapest, 1905–1906 Privatdozent univ. Kolozsvár (Austria-Hungary, today Cluj-Napoca/Romania), 1906–1911 senior assistant, 1911 extraord. prof., 1911–1959 prof. univ. Budapest.

[36] Comp., e. g., Pál Turán's introduction to the collected works of Fejér [Fej70, Bd. 1, p. 21–27].

$$\sigma_n(f,x) := \frac{\sum_{i=0}^{n-1} S_i(f,x)}{n}$$

are called Fejér sums of f.

Then Fejér's theorem is

Theorem 4.3 *Let f be a continuous 2π-periodic function. Then:*

$$\lim_{n\to\infty} \|f - \sigma_n(f,.)\| = 0.$$

The disadvantage of Fejér's summation is the fact that its partial sums do not interpolate the given continuous function and so it is not able to replace interpolation algorithms in general.

But Fejér could also solve this problem with the help of the today so-called *Hermite–Fejér Interpolation.*

In 1878 Hermite [Her78] defined a sequence of interpolating polynomials for given knots, which interpolated both the function and its derivative. Fejér studied (in [Fej16/1] and [Fej16/2]) a special case of those polynomials, where the zeros of the Chebyshev polynomials $x_1, \ldots, x_n \in [-1,1]$ were taken as knots for all $n \in \mathbb{N}$. Then define for $x \in [-1,1]$,

$$h_i(x) = [1 - 2(x - x_i)l_i'(x_i)] \left(l_i(x)\right)^2,$$

where the functions l_i are Lagrange's basic polynomials so that with the definition

$$\omega(x) := \prod_{j=1}^{n} (x - x_j)$$

there holds:

$$l_i(x) = \frac{\omega(x)}{\omega'(x_i)(x - x_i)}.$$

With the polynomials h_i it is easily possible to define the interpolation polynomial

$$H_n(x) := \sum_{i=1}^{n} f(x_i)h_i(x)$$

for an arbitrary function f.

Fejér's [Fej16/2] main result was:

Theorem 4.4 *Let $f \in C[-1,1]$. Then:*

$$\lim_{n\to\infty} \|f - H_n\|_\infty = 0.$$

4.6.2 Runge's Ideas about Approximation Theory

Gottfried Richenhagen [Ric85] clearly described in the scientific biography of Carl Runge[37] how his numerical mathematics were founded by a "pragmatic adoption of Weierstraß' analysis."

The subject of research in function theory in the sense of Weierstraß is a complex function which can be represented as a power series. It is well known that with this setting Weierstraß' theory differed from the function theory shaped by Cauchy in which the centre is set by Cauchy's integral formula.

Runge's first contribution to approximation theory was [Run85/1], where he proved the approximation theorem named by him.

The statement of this theorem is that *any function which has only removable singularities or poles in a certain domain can be arbitrarily well approximated by a uniformly convergent series of rational functions*. It has a similar meaning for complex approximation theory, like Weierstraß' theorem for real-valued approximation theory.

Another work from the same year [Run85/2] would have got more attention, if it had not been published nearly simultaneously with Weierstraß' approximation theorem: Runge showed that any continuous function can be approximated uniformly by *rational* functions. The proof was more elementary than that of Weierstraß –Runge approximated continuous functions by pieces of polygons, which he approximated then by rational functions, but of course the statement was weaker!

His concept of approximation theory became evident later, after he had worked on some of the important problems of interpolation and approximation theory which we have already talked about [Run01]. In 1904 his textbook "Theory and practice of series" [Run04] was published.

4.6.2.1 Approximation by Representation

As the title said, in his textbook Runge discussed the representation of functions by series expansions. He introduced it with a section about the 'concept of approximation,' where it became clear that not theoretical questions, but numerical approximations of functions which were easy to compute were important for him.

There Runge considered the well-known cases of power and Fourier series and added to them some of his own observations about interpolating series and series of Chebyshev polynomials. The characterizing bond of this work Richenhagen analysed as follows:

"Here a concept is introduced and expanded which can be accurately characterized by the catch-phrase 'Approximation by representation'

---

[37] Carl David Tolmé Runge (*Bremen 1856, †Göttingen 1927), Studies in München and Berlin, 1883–1886 Privatdozent in Berlin, 1886–1904 o. Prof. TH Hannover, 1904–1925 o. Prof. in Göttingen.

and which not only summarized old, already well-known approximation algorithms under a common leading aspect, but also made possible to derive new algorithms."[38]

Runge's concept of representation theoretically followed that of Weierstraß and extended it.

Weierstraß' basic concept, the *uniform convergence* of function series was also used like a property which also was due to Weierstraß, the *decomposition property*.

So a function f (real- or complex-valued) is approximated by a sequence of functions R_n, for which holds:

- $\lim_{n \to \infty} \|f - R_n\|_\infty = 0$ (uniform convergence) and
- $R(z) := \sum_{i=1}^{n} c_{n,i}(f) T_{n,i}(z)$ (Decomposition into parts, which depend on f or are independent of it).

All series expansions we have already discussed satisfy these conditions.

Theoretically less clear are the other properties Runge demanded for the basic functions $T_{n,i}$ and the coefficients $c_{n,i}$. Richenhagen pointed out that *explicitness and simplicity* were the most important features of Runge's approximation theory and gave examples like the choice of monomials or simple fractions like $\frac{1}{z_{n,i}}$ with given knots $z_{n,i}$ for $T_{n,i}$ and e. g., Taylor or Lagrange coefficients as $c_{n,i}$. These claims did not result from theoretical, but from practical considerations, since the number of computations was a decisive criterion here. So 'theory and practice of series' is less a theoretical work, where for example the problem of approximability of continuous functions by interpolatory polynomials, which was still unsolved then, was not discussed.

4.6.2.2 Expansion after Chebyshev Polynomials

An idea to compute approximation formulae often applied by Runge was the search for an approximation of the expression $\frac{1}{w-z}$, which is part of Cauchy's integral formula for functions f which are analytic on the disc $K_\rho(z_0)$ with radius $\rho < |w|$

$$f(z) = \frac{1}{2\pi} \int_{K_\rho(z_0)} \frac{f(w)}{w - z} \, dw.$$

If we now assume that $f(t)$ is analytical on a torus $A_{a,\frac{1}{a}}$, bounded by the radius a and $\frac{1}{a}$, then we reach by the transformation $t \mapsto \frac{t+t^{-1}}{2} =: z$ that $f(t)$

is analytical within the ellipse with focus ± 1 and with a and $\frac{1}{a}$ as the radius of the half-axes.

By expansions into power series Runge showed that within this ellipse for $w =: \frac{\tau + \tau^{-1}}{2}$ there holds

$$\frac{1}{w-z} = \frac{1}{w-\tau} \sum_{r=0}^{n-1} \tau^r (t^r + t^{-r}) + \frac{\tau^n}{w-\tau} \left(\frac{t^n}{1-\tau t} + \frac{t^{-n}}{1-\tau t^{-1}} \right),$$

where the error term uniformly converges to zero, if t lies within the torus $A_{a,\frac{1}{a}}$, z lies within the above-defined ellipse and w and τ lie beyond the respective sets.

The approximate representation is also a polynomial in z because replacing

$$Z_r(t) := \frac{t^r + t^{-r}}{2}$$

we have after transformation and using the binomial formula:

$$Z_r(z) := \sum_{j=0}^{\left[\frac{r}{2}\right]} \binom{r}{2j} z^{r-2j} \left(z^2 - 1\right)^j,$$

and so $\frac{1}{w-z}$ can be approximated by a uniformly convergent polynomial series, whose domain of convergence contains the interval $[-1, 1]$ which is bounded by the focus of the ellipse.

If we assume that the function to be approximated, f, is real-valued, then these results can be transferred.

Finally we get after the necessary transformations the functions Z_r in the representation

$$Z_r(x) = \cos r \arccos(x),$$

and so we get a representation in series of Chebyshev polynomials!

4.6.2.3 A Special Concept for Applied Mathematics—'Sensible Functions'

How Runge regarded the rôle of applied mathematics within the whole of mathematics, he would later explicitly describe in his book 'Graphical Methods' [Run15]:

"The solution of many, maybe all mathematical problems consists of the determination of the values of unknown quantities satisfying certain given conditions. It decays into several steps, whose first is the investigation whether the searched quantities exist so that it is possible to satisfy the given conditions or not. If the proof of the impossibility is made, then we are ready [...].

In many cases the first step to the solution gives so little difficulties that we can immediately pass to the second step, the research for a method to compute the unknown quantities. Or it may be suitable to begin with the second step even if the first is not so easy, since if it is possible to find methods to compute the unknown quantities the proof of their existence is done. But if it is not possible, then there is still time to return to the firstly mentioned step." [39]

Here his constructive approach became clear: The computation of unknown quantities is more important than the isolated proof of their existence. The existence proof is an interim solution, which makes sense only if we are not able to get further results.

What does it mean to compute unknown quantities? Of course this question was theoretically answered, if an arithmetic expression, a representation for the unknown quantity could be found. Runge, however, was not content with such an answer, since it is possible that such an expression is not suitable for practice, the expense to compute it might be too high. That 'not a little number' of mathematicians thought that their task was done with the solution of the above-described questions, Runge explained as follows:

"I believe that this is caused by the fact that the pure mathematician is not used to expanding his investigations to the reality. He leaves this to the astronomer, the physicist, the engineer. But these are interested only in the real numerical values coming from the mathematical computations. They are forced to execute the computations and when they do so they stand before the question if it might not be possible to reach the results on a shorter way or with less expenses. If we assumed that a mathematician would give them an absolutely sharp and logical method demanding 200 years of permanent computations they would

---

[39] The text is cited after the introduction of the second edition of [Run15]. Runge himself wrote that this edition remained mainly unchanged. Indeed this text is equal to the text of the first edition cited by Richenhagen [Ric85, p. 136 f.]: „Die Lösung vieler, wenn nicht aller, mathematischen Probleme besteht in dem Ermitteln der Werte unbekannter Größen, die gewissen gegebenen Bedingungen genügen. Sie zerfällt in verschiedene Schritte, deren erster die Untersuchung ist, ob die gesuchten Größen wirklich existieren, so daß es möglich ist, den gegebenen Bedingungen zu genügen, oder nicht. Ist der Beweis der Unmöglichkeit erbracht, so ist man mit dem Problem fertig [...].

In vielen Fällen kann der erste Schritt zur Lösung so wenig Schwierigkeiten bieten, daß man sofort zu dem zweiten, dem Aufsuchen der Methode zur Berechnung der gesuchten unbekannten Größen, übergehen kann. Oder es kann, selbst wenn der erste Schritt nicht so leicht ist, zweckmäßig sein mit dem zu zweit genannten anzufangen, denn wenn es gelingt Berechnungsmethoden zu finden, welche die unbekannten Größen bestimmen, so ist der Beweis ihrer Existenz ja einbegriffen. Gelingt es aber nicht, so ist es immer noch Zeit, zum erstgenannten Schritt zurückzukehren."

be right to regard this as little better than nothing. So there is a third step to the complete solution of a mathematical problem, namely to find the method which leads to the solution with the least expense of time and labor. I think that this step is a chapter of mathematics and is as good as the first two and that it must not be allowed to leave it to astronomers, physicists, engineers or others whoever wants to apply mathematical methods, since these people give their attention only to their results and tend to neglect the generalization of the methods they found, whereas in the hand of a mathematician these methods are developed from a higher point of view and the question of applicability to other problems, also from other subjects of scientific research is also taken into account."[40]

In his representation theory this opinion could be found in the claim for *explicitness and simplicity* of the coefficients and the approximating function. His pragmatic adoption of Weierstraß' approximation theory also contained a theoretical restriction, the demand for the (multiple) differentiability both of the function to be approximated and the series representing it.

Although analytical functions stand in the centre of Weierstraß' theory of functions, he explicitly referred to series expansion representing non-differentiable functions.

An example cited by Runge [Run04, p. 81 f.] was the function

$$f(x) := 1 + \sum_{i=2}^{\infty} \frac{\cos i^2 x}{i^2},$$

---

[40] [Run15, 2nd edition, p. 2:] „Dies beruht, glaube ich, auf der Tatsache, daß der reine Mathematiker nicht gewohnt ist, seine Untersuchungen auf die Wirklichkeit auszudehnen. Das überläßt er dem Astronomen, dem Physiker, dem Ingenieur. Diese wiederum interessieren sich hauptsächlich für die wirklichen numerischen Werte, die sich aus den mathematischen Berechnungen ergeben. Sie sind gezwungen, die Berechnungen auszuführen, und indem sie dies tun, werden sie vor die Frage gestellt, ob sich dasselbe Ergebnis nicht auf kürzerem Wege oder mit geringerer Mühe erreichen liee. Gesetzt der Mathematiker gibt ihnen eine zwar vollkommen scharfe und logische Methode an, die über 200 Jahre unausgesetzter Rechenarbeit zu ihrer Durchführung erfordert, so wären sie wohl berechtigt, dies für wenig besser als nichts anzusehen. So ergibt sich also ein dritter Schritt zur vollständigen Lösung eines mathematischen Problems, nämlich der, diejenige Methode zu finden, die mit dem geringsten Aufwand von Zeit und Mühe zur Lösung führt. Ich behaupte, da dieser Schritt gerade so gut ein Kapitel der Mathematik bildet, wie die beiden ersten und da es nicht angeht, ihn den Astronomen, Physikern, Ingenieuren und wer sonst noch mathematische Methoden anwendet, zu überlassen, weil diese Leute ihr Augenmerk nur auf die Ergebnisse richten, und daher geneigt sind, die Verallgemeinerung der von ihnen etwa ersonnenen Methoden zu vernachlässigen, wogegen in der Hand des Mathematikers die Methoden von einem höheren Gesichtspunkte aus entwickelt werden und die Frage nach ihrer Anwendbarkeit auf andere Probleme, auch solche anderer Gebiete wissenschaftlicher Forschung, gehörige Berücksichtigung findet."

which is continuous everywhere because this series uniformly converges everywhere. Differentiating it we see that there is no interval where the series of derivatives uniformly converges. Therefore f is not continuously differentiable.
Runge wrote about this:

> "Firstly it was shown by Weierstraß that functions can be represented by such series which do not have a differential quotient. This might be very instructive for the development of mathematical concepts, but we can say that for the practical application of mathematics to empirical problems such functions and such representations do not make sense."[41]

It is also interesting that by a footnote Runge explicitly mentioned Felix Klein's lectures 'Application of Differential and Integral Calculus to Geometry, a Revision of the Principles' from 1902 and so accepted the concept of 'sensible' functions.[42]

Consequently in the 'Theory and Practice of Series' Runge dealt only with such series expansions.

Clearly we see a similarity of the principles of the St Petersburg Mathematical School and Runge's conception. For Runge application also took the leading rôle and he always tried to have specific problems in mind in his later mathematical work. The aim of solving a problem to the "receipt of a suitable formula or a good algorithm being appropriate for practical computations"[43] matched with Runge's ideal. We want to remember that Chebyshev and his pupils did not prove any existence theorem which is not a trivial corollary of the explicit determination of a solution of the regarded problem.

But Carl Runge's scientific curriculum vitae showed that he arrived at his ideas about applied mathematics from a long way off starting from pure mathematics and a deep knowledge of the methods of Weierstraß' function theory. He also was a supporter of Weierstraß' 'constructive view to subjects' [Ric85] and did not attach much value to descriptive theories dealing with mathematical objects independent from their concrete representations.

---

[41] [Run04, p. 82]: „Es ist zuerst von Weierstraß gezeigt worden, da durch solche Reihen Funktionen dargestellt werden können, die gar keinen Differentialquotienten besitzen. So lehrreich dies nun auch für die Begriffsentwicklung ist, so kann man doch sagen, da für die praktische Anwendung der Mathematik auf empirische Probleme solche Funktionen und solche Darstellungen keine Bedeutung haben."

[42] Weierstraß'"monster' caused Felix Klein to introduce some claims of differentiability to reach a better handling of empirically given functions. He thought that the demand for (multiple) differentiability is fundamental for empirical functions. Without a clear definition he called such functions 'vernünftig' ('sensible'). We don't want to discuss this subject extensively because this would lead far beyond the borders of this work. We only want to refer to [Kle28], the third edition of the above-mentioned lectures.

[43] Compare [Ozhi66, S. 61].

The approaches of St Petersburg mathematicians mainly dealing with the solution of extremal problems and searching for functions satisfying certain side-conditions and not for their representations were more of a descriptive nature. We saw that their common link was the search for an approximative expression converging for all analytical functions and having the form

$$T_n(f) = T_n(\sigma_n, \sigma_{n+1}, \dots)$$

with

$$f(x) := \sum_{i=0}^{\infty} \sigma_i x^i.$$

The series expansion coming out now is

$$f(x) \sim T_0(f) + \sum_{k=0}^{\infty} T_{k+1}(f) - T_k(f)$$

and would not satisfy the decomposition property demanded by Runge (and Weierstraß), because the T_k just depend also on the coefficients $\sigma_k, \dots$. Obviously the conception of best approximation contradicted Runge's conception, not only by the approach, but also in the aim.

Maybe the St Petersburg mathematicians would agree with the concept of a 'sensible' function,[44] but there is not known anything about discussions on this concept, which is said to be used firstly by Jacobi.[45] But in St Petersburg discussions on such topics were not made very often, as we came to know.

Besides this Runge is not as prejudiced in the choice of mathematical methods as his St Petersburg colleagues. He accepted both Weierstraß' and Cauchy's approach in function theory and also was interested in Riemann's work[46]—the latter approaches both were more or less clearly rejected by Chebyshev, Korkin and Lyapunov.

So the bridge between Chebyshev and Runge can be built because of their orientation to practice and the resulting common opinion about the subject to be investigated, but according to both methods and aims, their approaches differed.

4.7 Quantitative Approximation Theory

It has already been emphasized several times that Weierstraß' approximation theorem marked a milestone in approximation theory. He gave a positive answer to the question whether for any function $f \in C[a, b]$ the quantity

---

[44] We want to remember Chebyshev's opinion about 'philosophizing' in mathematics, the implicit assumptions of differentiability and Posse's lectures [Pos03], where he rejected Weierstraß' monster as a subject of his studies.

[45] Comp. [Kle28, p. 50].

[46] You find more details about Runge's opinion about function theory in [Ric85, chapter II].

$$E_n(f) = \min_{p \in P_n} \|f - p\|_\infty$$

converges for for $n \to \infty$ to zero.[47]

Based on this result other questions suggest themselves:

1. How fast can a given continuous function be approximated?
2. For which functions does $nE_n(f)$ converge?
3. Which polynomial sequences allow a fast convergence?

Quantitative approximation theory deals with such questions.

In the overview [Fis78] Stephen D. Fisher summarized its aim in the following way (p. 318):

"Quantitative Approximation Theory attempts to determine as precisely as possible the size of the error in this approximation given specific information about the function to be approximated and the set of functions from which the approximant is to be taken."

The first contributions to this subject were written by Henri Lebesgue[48] and Charles-Jean de La Vallée-Poussin.[49] For some special cases they could determine the order of convergence.

We want to put their results together in the following table:

Citation	Assumption	Order of Convergence
[Leb08]	$f \in \mathrm{Lip}_1$	$O\left(\sqrt{\frac{\log n}{n}}\right)$
[Val08/1]	$f \in \mathrm{Lip}_1$	$O\left(\frac{1}{\sqrt{n}}\right)$
[Val08/2]	$f' \in \mathrm{BV}$	$O\left(\frac{1}{n}\right)$
[Leb10/1]	$f \in \mathrm{DL}$	$o\left(\frac{1}{\log n}\right)$
[Leb10/2]	$f \in \mathrm{Lip}_1(2\pi)$	$O\left(\frac{\log n}{n}\right)$

To explain these results we want to remember some of the assumptions:

- The Lipschitz condition ($\alpha > 0$):
  $f \in \mathrm{Lip}_\alpha([a,b]) \iff \exists L \in \mathbb{R}$ so that for all $x, y \in [a,b] : |f(x) - f(y)| \le L|x - y|^\alpha$.

---

[47] Weierstraß also showed that for any 2π-periodic function there holds the analogous statement involving trigonometric polynomials. We want to concentrate on the case of algebraic polynomials, because nearly all results here have an analogy there.

[48] Henri Léon Lebesgue (*Beauvais (France) 1875, †Paris 1941), studies at the École Normale Supérieure, 1899–1910 in Nancy, 1902 doctorate univ. Nancy, 1910–1941 prof. at the Sorbonne.

[49] Charles-Jean Gustave Nicolas de La Vallée-Poussin (*Louvain/Leuven (Belgium) 1866, †Louvain 1962), studies of engineering sciences and mathematics, 1891–1893 assistant univ. Louvain, 1893–1943 ord. prof. univ. Louvain, 1909–1966 ord. member of the Belgian academy of sciences.

- The Dini–Lipschitz condition:
  $f \in \mathrm{DL} \iff \lim_{\delta \to 0} \omega(f, \delta) \log \delta = 0$, where

  $$\omega(f, \delta) = \max_{|x-y| \leq \delta} |f(x) - f(y)|$$

  is the first modulus of continuity.[50]
- Functions of bounded variation:
  $f \in \mathrm{BV}[a, b] \iff \exists M \in \mathbb{R}$ so that for all decompositions $z : a = t_0 < t_1 < \cdots < t_n = b$ of the interval $[a, b]$ there holds:

  $$\sum_{i=1}^{n} |f(t_i) - f(t_{i-1})| \leq M.$$

We see in the case of the Lipschitz condition how the results were permanently improved. But we have to remark that Lebesgue's last result explicitly only holds for 2π-periodical functions.

The first work which gave a theoretical framework to these results and could improve them enormously was the doctoral thesis of Dunham Jackson[51] "About the Exactness of the Approximation of Continuous Functions by Polynomials of Given Degree and Trigonometric Sums of Given Order" [52] written in 1911 and acknowledged with the prize of the Göttingen faculty.

4.8 Jackson's Thesis

As a postgraduate supported by Harvard university, Jackson studied for four semesters in Germany (1909–1911), the first three of them in Göttingen, the last in Bonn.[53]

Presumably Jackson was helped to arrange this trip by his teachers[54]

---

[50] Lebesgue was the first (in [Leb10/1]) to use the name ω, but not to formulate the Dini–Lipschitz condition.

[51] Dunham Jackson (*Bridgewater, Mass. (USA) 1888, †Minneapolis (USA) 1946), 1904–1909 studies at Harvard univ. in Cambridge (Mass.), 1909–1911 studies in Göttingen and Bonn, 1911–1916 instructor Harvard univ., 1916 assistant prof., 1919–1946 full prof. univ. of Minnesota, Minneapolis.

[52] The original name was „Über die Genauigkeit der Annäherung stetiger Funktionen durch ganze rationale Funktionen gegebenen Grades und trigonometrische Summen gegebener Ordnung" [Jack11]

[53] Compare [Har48] and [Amt10].

[54] Jackson mentioned them, amongst other influences, in his curriculum vitae added to his dissertation.

Bôcher[55] and Osgood,[56] who both did postgraduate studies in Gôttingen.

His academic teacher there was Edmund Landau, who presented him with three different subjects to be chosen among for his thesis—Jackson decided to write about a topic which had been formulated as a praiseworthy problem of the faculty as follows:[57]

> "It is well-known that 25 years ago Weierstraß proved firstly that any function continuous in an interval can be approximated as precisely as possible by a polynomial. Recently by de la Vallée Poussin [...] and Lebesgue there have been made first investigations about the dependence between the necessary smallest possible degree of the approximating polynomial and the prescribed boundary of exactness.[58] Whether these estimates of the degree as a function of the exactness can be improved is an open set of problems.
>
> The faculty wishes that there will be a fundamental progress in this direction; such a progress would lie in the answer of the following question asked by de la Vallée Poussin (p. 403):[59] Does the product of boundary of exactness and respective minimal degree converge to zero in the case of a fixed given polygonal line?"[60]

The expectations of the faculty were fulfilled, Jackson's results gave the sought answer to some fundamental questions of quantitative approximation theory. Only a small drop of bitterness remained, since shortly before the printing of the thesis there was published a contribution by Sergey Natanovich

---

[55] Maxime Bôcher (*Boston 1867, †Cambridge (Mass., USA) 1918), 1889/90 studies in Gôttingen, taught at Harvard from 1894 until 1918. His subjects of research were differential equations, series and algebra.

[56] William Fogg Osgood (*Boston 1864, †Belmont (Mass., USA) 1943), 1887–1889 studies in Gôttingen and Erlangen, taught at Harvard from 1890 until his retirement. His subjects were function theory and variational calculus.

[57] The posed problem is a part of the published thesis [Jack11].

[58] Compare the table on page 159.

[59] They had in mind [Val08/2].

[60] „Bekanntlich hat Weierstraß vor 25 Jahren zuerst bewiesen, da jede in einem Intervall stetige Funktion mit beliebiger Genauigkeit durch eine ganze rationale Funktion approximiert werden kann. über die Abhängigkeit des hierzu erforderlichen kleinstmöglichen Grades dieses Polynoms von der vorgeschriebenen Genauigkeitsgrenze sind die ersten Untersuchungen in neuerer Zeit gemacht worden, von de la Vallée Poussin [...] und Lebesgue [...] Ob die hierbei erzielten Abschätzungen des Grades als Funktion der Genauigkeitsgrenze noch übertroffen werden können, ist ein offener Fragenkomplex.
   Die Fakultät wünscht, da in dieser Richtung ein wesentlicher Fortschritt gemacht werde; ein solcher würde z. B. in der Beantwortung der folgenden von de la Vallée Poussin (S. 403) gestellten Frage liegen: Konvergiert im Falle eines festen gegebenen Linienzuges das Produkt von Genauigkeitsgrenze und zugehörigem Minimalgrad gegen Null?"

Bernstein, where some very similar results were contained. We will return to this fact in section 5.3.4.

4.8.1 Jackson's Theorem

Jackson's main result did not only improve the results of de la Vallée-Poussin and Lebesgue, but could also give an upper bound for the speed of polynomial approximation to continuous functions and in the case of periodic functions for trigonometric sums. The rôle of the modulus of continuity as a measure for the approximation speed became clear for the first time.

Theorem 4.5 (Jackson's Theorem) $\exists K \in \mathbb{R}$ *so that for all continuous functions* $f : [a, b] \to \mathbb{R}$ *and* $n \in N$ *there holds:*

$$E_n(f) \leq K\omega(f, \frac{b-a}{n}). \tag{4.5}$$

The proof of this theorem is based on an auxiliary theorem which alone would have already improved some of its preceding results: If f satisfies the Lipschitz condition $f \in \mathrm{Lip}_1[a, b]$, then:

$$E_n(f) \leq O\left(\frac{1}{n}\right).$$

This theorem is to be proved by elementary estimations of the functions

$$I_m(x) = \frac{k_m}{2} \int_a^b f(u) \left[\frac{\sin m(u-x)}{m(u-x)}\right]^4 du \quad \text{with}$$

$$\frac{1}{k_m} = \int_a^b \left[\frac{\sin mu}{mu}\right]^4 du.$$

They are used as first approximations of f and after this they are approximated themselves by polynomials.

De la Vallée-Poussin was able to determine his results with a similar method, namely he reached an approximation order of $O\left(\frac{1}{n}\right)$ for functions of bounded variation through the approximation of f by the integrals

$$J_m(x) = \frac{1}{\pi} \int_a^b f(u) \frac{\sin m(u-x)}{m(u-x)} du.$$

Presumably he so gave the impulse for Jackson's proof.

With this theorem an immediate derivation of the main result is possible by a tricky construction:

If the interval $[a, b]$ is divided into equidistant parts, $a = x_0 < x_1 < \cdots < x_n = b$, and if $\bar{f}$ is defined as that function which coincides with f in these knots and is continued linearly else, then $\bar{f}$ satisfies a Lipschitz condition, to be precise, then there holds for all $x, y \in [a, b]$:

$$|\bar{f}(x) - \bar{f}(x)| \leq \frac{\omega\left(\frac{b-a}{n}\right)}{\frac{b-a}{n}}|x - y|.$$

Thus, with the auxiliary theorem there holds for an arbitrary polynomial $p \in \mathbb{P}_n$ and an arbitrary $x \in [a, b]$:

$$|\bar{f}(x) - p(x)| \leq L\omega\left(\frac{b-a}{n}\right)$$

with a constant L, and with the definition of $\bar{f}$ we have

$$|f(x) - \bar{f}(x)| \leq 2\omega\left(\frac{b-a}{n}\right),$$

and so the searched for result has been proved.

Extensions of the auxiliary theorem lead to some interesting results for functions $f \in C^{k-1}[a, b]$, which additionally satisfy $f^{(k-1)} \in \mathrm{Lip}_1$.

For them there holds:

1. $E_n(f) = O\left(\frac{1}{n^k}\right)$,
2. $\forall n \in \mathbb{N} \; \exists p \in \mathbb{P}_n$, so that:

$$\|f - p\| = O\left(\frac{1}{n^k}\right)$$

$$\|f' - p'\| = O\left(\frac{1}{n^{k-1}}\right)$$

$$\vdots$$

$$\|f^{(k-1)} - p^{(k-1)}\| = O\left(\frac{1}{n}\right).$$

An analogous result for the trigonometric case was also proved by Jackson.

4.8.2 Further Results. An Inverse Theorem

Now the natural question is whether this approximation is the best possible. Or, if a function $\varphi(n)$ describes the order of approximation (that is, $E_n(f) = O(\varphi(n))$), then one has to show that for any function $\psi(n)$ also satisfying $E_n(f) = O(\psi(n))$ their quotient $\psi(n)/\varphi(n)$ does not converge to zero.

Jackson could not solve this problem completely, but anyhow he managed to show:

Theorem 4.6 *For any positive number η there exist functions $f \in \mathrm{Lip}_1[a, b]$, for which do not exist a constant K satisfying the inequality*

$$E_n(f) \leq K\frac{1}{n^{1+\eta}}.$$

He also found analogies for the case of functions with their $(k-1)^{st}$ derivative being Lipschitz continuous and for the trigonometric approximation.

It should be emphasized that Jackson also proved a so-called inverse theorem. Such a theorem is characterized by the problem to get information about the smoothness of the function to be approximated by its approximation order.

Theorem 4.7 (Inverse Theorem by Jackson) *Let $f : [a,b] \rightarrow \mathbb{R}$ be a function. If there exist a constant $K \in \mathbb{R}$ and a number $\eta > 0$ so that for all $n \in \mathbb{N}$ the inequality*

$$E_n(f) \leq K \frac{1}{n^{2k+\eta}}$$

holds, then:
f is $2k$-times continuously differentiable on (a,b).

An analogous theorem was also formulated for 2π-periodic functions.

At first he proved the trigonometric version of the theorem, where he used an auxiliary theorem which states a connection between the approximation order by general trigonometric sums and partial sums of the Fourier series—a kind of a quantified version of Fréchet's theorem. So he got an approximation which was easier to handle and found a series uniformly converging to the function f and satisfying the demanded differentiability property.

Looking at this theorem we see how strange the discussion is, led by the question of who has priority with respect to the fundamental theorems of quantitative approximation theory. Whereas no-one denies that the direct theorems are attributed to Jackson, on the other hand no one doubts that the inverse theorems belong to Bernstein, although both published similar direct and inverse results in their monographs [Jack11] and [Bern11] which were published nearly simultaneously in summer 1911 with a difference of about three weeks. We will return to this question in section 5.3.4, and try to illuminate it once again because of current events and doubt about the independence of their contributions.

Jackson's thesis closes with generalizations of the direct theorems to the case of higher-dimensional functions. Analogies for functions satisfying certain Lipschitz conditions are proved.

4.8.3 How the Faculty Judged

Jackson's thesis was a milestone of the development of modern approximation theory. It gave the foundations to investigate the quality of approximation algorithms and marked the starting point for a partial subject of approximations which gives new and ingenious results until modern times.

The philosophical faculty of Göttingen University saw their expectations fulfilled, It judged:[61]

---
[61] These words were also included in the published version of Jackson's thesis.

"[...]In the main part of his work the author went beyond the former borders of our knowledge in several aspects. His results should be rated very high, since Lebesgue and de la Vallée-Poussin, whom we have to thank for the origin and support of all these sets of problems, have published several further results in this direction in the meantime, that is, since the posing of the praiseworthy problem, but without reaching the results the author could prove.

Because of all these reasons the treatise has to be regarded as a very good dealing with a praiseworthy question. The author succesfully became familiar with a difficult and extensive subject and enriched science with valuable results in competition with mathematicians of the first rank. [...] Therefore the faculty awards the prize to the work."[62]

4.9 A Note About Göttingen's Rôle

The mathematical 'Mecca', as Göttingen university had often been called at the beginning of the 20th century with respect to outstanding mathematicians like David Hilbert and Felix Klein, left its traces in approximation theory too.

We saw that in the period from about 1900 (Fejér's work about the summability of functions [Fej00]) until 1911 (Jackson's thesis [Jack11]) an abundance of pioneering results came to light, theoretically satisfactorily solving several problems which had been open for a long time, laying the foundations for new mathematical subjects, underpinning and quantifying theoretical approaches and putting them in concrete terms.

David Hilbert's and Felix Klein's mediator rôle should not be underrated, since their school was sufficiently attractive for most of the above-mentioned mathematicians (we might say: for all except the French) to spend there one or more semesters as Table 4.1 shows.

The next chapter is devoted to the work of Psheborski and much more to the early contributions of Sergey Natanovich Bernstein. We will then see, how also Göttingen's influence was important for the confluence of Weierstraß' and Chebyshev's theoretical approaches to *Constructive Function Theory*.

---

[62] „[...] Im Hauptteil seiner Arbeit ist [der] Verfasser in mehreren Beziehungen über die bisherigen Grenzen des Wissens hinausgegangen. Seine Ergebnisse sind umso höher zu bewerten, als Lebesgue und de la Vallée-Poussin, denen die ganzen Fragestellungen ihre Entstehung und Förderung verdanken, inzwischen d. h. seit Stellung dieser Preisaufgabe weitere Ergebnisse in dieser Richtung publiziert haben, ohne einige bestimmte Ziele zu erreichen, zu denen [der] Verfasser gelangt.

Aus allen diesen Gründen ist die Abhandlung als eine sehr gute Bearbeitung der Preisfrage anzusehen. Der Verfasser hat sich in einen schwierigen und umfangreichen Stoff erfolgreich eingearbeitet und hat im Wettbewerb mit Mathematikern ersten Ranges die Wissenschaft um wertvolle Ergebnisse bereichert. [...] Daher erkennt die Fakultät der Arbeit den Preis zu."

	SS 01	WS 01	SS 02	WS 02	SS 03	WS 03	SS 04	WS 04	SS 05	WS 05
Klein	P	P	P	P	P	P	P	P	P	P
Hilbert	P	P	P	P	P	P	P	P	P	P
Kirchberger	S	S	S							
Fejér				S						
Bernstein					S	S	S			
Psheborski								S		
Runge								P	P	P
Haar										S

	SS 06	WS 06	SS 07	WS 07	SS 08	WS 08	SS 09	WS 09	SS 10	WS 10
Klein	P	P	P	P	P	P	P	P	P	P
Hilbert	P	P	P	P	P	P	P	P	P	P
Bernstein									S	
Runge	P	P	P	P	P	P	P	P	P	P
Haar	S	S	S	S	S	S	S	S	L	L
Jackson								S	S	S

Table 4.1. *Stays of mathematicians in Göttingen between summer semester (SS) 1901 and winter semester (WS) 1910. We marked studies (S), lecturer's job (L) and professor's job (P).*

5

Constructive Function Theory: Kharkiv

The development of Russian approximation theory was continued not in St Petersburg, but in Kharkiv. The theories spread in Western Europe and the work of the St Petersburg Mathematical School there fell onto a fertile field because of the activities of Antoni-Bonifatsi Pavlovich Psheborski and Sergey Natanovich Bernstein.

Kharkiv had already had a very good reputation as the mathematical centre of the Ukraine: In 1879 there had been founded one of the first mathematical societies of the Russian Empire and some outstanding mathematicians worked at Kharkiv university, e. g., Tikhomandritski (1883–1904), Lyapunov (1885–1902), Steklov[1] (Lecturer's job from 1891 until 1906) and Grave (1899–1901).

5.1 Antoni-Bonifatsi Pavlovich Psheborski

The rôle of the Polish-Ukrainian mathematician Antoni-Bonifatsi Pavlovich Psheborski was important in three aspects: as a mediator between Western European and Eastern European ideas, as an author of contributions to approximation theory and as a supporter of Sergey Bernstein.

---

[1] Vladimir Andreevich Steklov (*Nizhni Novgorod 1864, †Gaspra (Crimea) 1926), studies in Moscow and Kharkiv, 1887 candidate-thesis (under supervision of Lyapunov), 1893 master thesis about the motion equations of a solid body in liquids, joint research with Lyapunov, 1891–1896 privatdotsent univ. Kharkiv, 1896–1902 extraord. prof. univ. Kharkiv, 1902–1906 ord. prof. univ. Kharkiv, since 1906 ord. prof. univ. St Petersburg, 1910 ordinary member of the academy of sciences, 1919 its vice president. Steklov had an outstanding reputation as science organisor.

5.1.1 His Biography

Antoni-Bonifatsi Pavlovich Psheborski was born May 14, 1871 as a son of
the Polish naval doctor Pavel Antonovich Psheborski and his wife Marina
Milenovskaya in the village of Khoroshee, district of Lipovets in the Kyiv
province (now Vinnitsa province).[2]

After his school education at the classical Aleksandrovski grammar school
in Nikolaev which he finished with the gold medal in 1889, he registered at the
physico-mathematical faculty of the university of the holy Vladimir in Kyiv.
His teachers there were among others M. E. Vashchenko-Zakharchenko,[3] V. P.
Ermakov,[4] G. K. Suslov[5] and P. M. Pokrovski,[6] who would have the largest
influence in Psheborski's education. In 1894 he finished the university education with the 'diploma of first stage' and the same year he won the gold medal
in a student's competition with his work 'Explanation of Weierstraß' method
within the theory of elliptic functions and construction of a link between the
concepts of Weierstraß and Jacobis. Afterwards he remained at the university
to prepare for the post of a professor.

With Pokrovski's supervision he mainly dealt with the theory of higher
transcendental functions based on the work of Legendre, Abel, Jacobi, Riemann und Weierstraß. He was awarded a prize for the work 'About methods
of Abel, Jacobi, Liouville and Weierstraß in the theory of elliptical functions.'
His Kyiv period of activity exclusively concentrated on this branch of function
theory.

In 1897 he took and passed two oral exams to get the venia legendi at Kyiv
university. But it was rejected because of his Polish origin and his belonging
to the Catholic church, although Pokrovski personally supported his cause
several times.

---

[2] We took the biographical details from [But92].

[3] Mikhail Egorovich Vashchenko-Zakharchenko (1825–1912), studies in Kyiv and
Paris (1847–48), he finished his studies in Kazan 1854 (candidate), 1862 master
thesis. 1855-1863 Lecturer at the Kyiv cadet school, 1863–1867 privatdotsent
univ. Kyiv, 1867 extraord., 1868 ord. prof. univ. Kyiv. Contributions to geometry,
function theory and history of mathematics.

[4] Vasili Petrovich Ermakov (1845–1922), studies in Kyiv until 1868, 1868–1874
privatdotsent, 1874–1879 lecturer, since 1879 ord. prof. univ. Kyiv, since 1884
corresponding member of the academy of sciences. Worked on variational calculus.

[5] Gavril Konstantinovich Suslov (1857-?), studies at St. Petersburg univ. until 1880,
master thesis there 1888. 1888-1891 extraord. prof. univ. Kyiv, since 1891 ord.
prof. univ. Kyiv. Worked on theoretical mechanics.

[6] Petr Mikhaylovich Pokrovski (1857–1901), studies in Moscow until 1881, 1883–
1885 teacher of mathematics at Moscow grammar schools, 1885–1889 Privatdotsent univ. Moscow, 1889-1890 studies in Berlin at Weierstraß, 1891 extraord.
prof. univ. Kyiv, 1894 ord. prof. univ. Worked on theory of functions, especially
elliptical and ultraelliptical functions.

Then in 1898 Psheborski moved to Kharkiv, where he became ordinary lecturer at the technological institute. In 1899 he additionally taught as a privat-dotsent at Kharkiv university.

In the Kharkiv period his mathematical subjects of interest were extended: Stimulated by Steklov and Grave he now was mainly engaged in questions from differential geometry, variational calculus, theoretical mechanics and (later) from approximation theory.

In October 1902 he defended his master thesis 'Some applications from the theory of linear congruences' before the opponents B. K. Mlodzeevski and K. A. Andreev[7] in Moscow.

5.1.2 Psheborski in Göttingen

His interest in variational calculus led him to Göttingen in summer semester 1904, where he could exchange ideas, especially with Hilbert. Unfortunately for him, Hilbert was engaged in a fundamental reorganisation of his lectures so he was unable to attempt all of them.[8] And so he only could visit Hilbert's lectures about function theory, Klein's lectures about differential equations and 'linear and spherical geometry' of Minkowski.

So Psheborski used his stay to become familiar with the pedagogical peculiarities of the lecturers, the organization of the mathematical education and the scientific atmosphere in Göttingen.

Here he was especially impressed by the manner of "such an outstanding teacher like professor Klein."[9]

He was enthusiastic about the enormous scientific freedom in Göttingen, which he observed was realized in fewer duties for students to pass examinations and in the bureaucracy which was pleasantly less at Göttingen university. About this he judged summarizing:

> "And so we see that the whole education at Göttingen university forms a slender system, whose aim is to give the possibility to learn to work independently. All institutions support this aim, all lecturers work in this direction. And they try to reach this aim without counting caps or formal registration of the students visiting a lecture, without obligatory controls or examinations; all are working voluntarily, lecturers and auditors are working together, in the relations between them there rules complete trust and veneration. No university administration is

---

[7] Konstantin Alekseevich Andreev (*Moscow 1848, †Moscow 1921), studies in Moscow, since 1873 privatdotsent, 1880-1898 ord. prof. univ. Kharkiv, 1898-1921 ord. prof. univ. Moscow, since 1884 corresponding member of the academy of sciences, 1879 foundation president of the Kharkiv Mathematical Society. He worked about projective geometry and analysis.

[8] For more detailed information about Psheborski's stay in Göttingen compare his report printed in 1906 [Psh06].

[9] [Psh06, p. 27]: «такого выдающегося педагога, как проф. Klein ».

butting in, whose rôle is determined by the observation of the external order; for this aim three officials are sufficient who form the whole staff of employees of the huge auditorial building. How far is that from our rules!"[10]

His enthusiasm about Göttingen university led him to the following statement, which was politically incorrect in a triple sense:

"The complete trust and the veneration that the teaching staff enjoys in the eyes of the students are a consequence of the conviction that an independent [...] professoriate that stands without pressure from outside can take between them only a worthy member, a person who devotes its power to the greatest deity—the science whose veneration made Germany one of the most cultivated states."[11]

After his stay in Göttingen Psheborski visited the third international congress of mathematicians[12] in Heidelberg, where he was especially impressed by Hilbert's talk about integral equations and Voronoy's talk about quadratic forms. Possibly there he got to know S. N. Bernstein, at least it is probably the first time that they both were at the same place at the same time. Bernstein had already left Göttingen when Psheborski arrived there.[13]

---

[10] [Psh06, p. 31]: «Итак, мы видим, что все преподавание в Геттингене представляет одну стройную систему, цель которой дать возможность всякому желающему научиться самостоятельно работать. Все учреждения способствуют этой цели, все преподаватели работают в этом направлении. И поставленная цель достигается без всякого счета шапок и записывания посещающих лекции студентов, без обязательных контролей и экзаменов; все занимаются добровольно, преподаватели и слушатели работают сообща, в отношениях между теми и другими царит полное доверие и уважение. В эти отношения не вмешивается никакая университетская администрация, вся роль которой сводится к наблюдению за внешним порядком; для последней цели достаточно трех служащих, составляющих весь штат служителей в громадном зале аудиторий.»

[11] [Psh06, p. 31 f.]: «Полное доверие и уважение, которыми пользуется профессорская коллегия в глазах студентов, являются следствием убеждения, что самостоятельная [...] коллегия, не находящаяся ни под чим давлением извне, может избрать в свою среду только лицо достойное, лицо, посвящающее свои силы величайшему божеству - науке, уважение к которой сделало Германию одним из культурнейших государств мира.» We want to emphasize that this statement was democratic, antipatriotic and blasphemous, and so it is clear the report was printed only after the 1905 revolution. Psheborski expressed his regret to that in a footnote to the printed report.

[12] Compare [Ver04] and again [Psh06].

[13] Compare Table 4.1 on page 166, [Amt10] and [Stef99].

5.1.3 Doctoral Thesis and Lecturer's Job

In 1905 Psheborski became extraordinary professor at Kharkiv university and had to give up his paid job at the technological institute, but continued to give lectures there.

He taught calculus, analytical geometry, variational calculus and numerical mathematics.

In 1908, again in Moscow, Psheborski defended his doctoral thesis 'Investigations about the theory of analytical functions and the continuation of Taylor's series'. Here his opponents were D. F. Egorov[14] and L. K. Lakhtin.[15] He then returned to his old subjects of research from Kyiv times.

The Russian revolution of October 1917 firstly had no influence on Psheborskis pedagogical activities; he remained professor at Kharkiv university and even was elected its rector in 1919. After the reorganisation of the university and the creation of the 'academy of theoretical sciences' (Академия теоретических знаний) replacing the university, he became its rector and at the same time head of the chair of theoretical mechanics. After a short while Psheborski was arrested because of an alleged espionage for Poland, but after 20 days of arrest he was freed and returned to the academy as dean of the physico-mathematical faculty, that is, degraded. In 1921 the academy was reorganized and renamed 'Institute for People's Education' (инситут народного образования). Psheborski became rector of this institute.

In July of the same year Psheborski consulted the Polish minister of education and asked for the possibility to move to Poland. Simultaneously he was offered an appointment as ordinary professor of the chair of mathematics at the university of Vilnius.[16] Because of the cholera and typhoid epidemic, however, he was not allowed to leave the Ukraine. So he moved to Poland only in 1922, at first for a short time to Vilnius, afterwards to Warsaw, where he became head of the chair of mechanics of the university. There he helped build his own institute for theoretical mechanics, hydrodynamics and 'motion of solid bodies and liquids'.

Besides this he taught analytical geometry, calculus and differential equations at the Warsaw polytechnical institute.

---

[14] Dmitri Fedorovich Egorov (1869–1931), since 1913 ord. prof. univ. Moscow, 1922–1930 president of the Moscow Mathematical Society, 1929 honorary member of the Soviet academy of sciences; in 1930 he was suspended from the Moscow Mathematical Society, removed from all positions, arrested and exiled to Kazan. Together with his pupil Nikolay Nikolaevich Luzin (1883–1950) Egorov was the founder of the theory of real-valued functions. For a good overview about the Moscow Mathematical school of Egorov and Luzin see [Pau97].

[15] Leonid Kuzmich Lakhtin (1863–1927), 1892–1896 ord. prof. univ. Tartu (Estonia, Russian name: Yurev), 1896–1927 ord. prof. univ. Moscow, worked on differential equations and mathematical statistics.

[16] Polish: Wilno.

In the Warsaw period his main subjects of research were theoretical mechanics and variational calculus.

He remained at the university until his death. Antoni-Bonifatsi Pavlovich Psheborski died in Warsaw May 24, 1941.

5.1.4 An Extension of V. A. Markov's Problem

Twenty years after the last St Petersburg contribution to the Chebyshev theory, Psheborski's work 'About some polynomials least deviating from zero on a given interval' was published[17] [Psh13/2]. It was a classical contribution, since it solved a minimization problem of Chebyshev type.

The problem generalized V. A. Markov's problem. Psheborski considered this case for *two* given side-conditions of the coefficients:

Minimize $\|p\|$, *where*

$$p(x) = \sum_{i=0}^{n} a_i x^i \in \mathbb{P}_n, \quad x \in [a, b]$$

and the coefficients of p satisfy the linear equations

$$\alpha = \sum_{i=0}^{n} \alpha_i a_i$$

$$\beta = \sum_{i=0}^{n} \beta_i a_i$$

with given real numbers α, $\alpha_0, \ldots, \alpha_n$, β *and* $\beta_0, \ldots, \beta_n$.

We want to use Psheborski's abbreviations and write for $p \in \mathbb{P}_n$:

$$\omega(p) := \sum_{i=0}^{n} \alpha_i a_i$$

$$\omega_1(p) := \sum_{i=0}^{n} \beta_i a_i.$$

But he added the following restriction: there should exist two different indices i and k for which there holds: both determinants

$$\begin{vmatrix} \alpha_i & \alpha_k \\ \beta_i & \beta_k \end{vmatrix} \qquad \begin{vmatrix} \alpha & \alpha_k \\ \beta & \beta_k \end{vmatrix}$$

---

[17] The only Russian papers belonging to this subject and being published in the time in between had very special subjects, as their titles already tell. They are [Sve01] and [Tra92] (cited after [But92]).

are not equal to zero. Therefore the problem did not generalize V. A. Markov's problem[18], because if there is only one side-condition, we have $\beta_0 = \cdots = \beta_n = \beta = 0$, and both determinants vanish. But it is a genuine generalization of Zolotarev's problem (compare section 3.2.2), since two given coefficients lead to the equations

$$a_n = 1 \quad \text{and} \quad a_{n-1} = \sigma,$$

and choosing n and $n - 1$ as the regarded indices the determinants become

$$\begin{vmatrix} 1 & 0 \\ 0 & 1 \end{vmatrix} \quad \text{and} \quad \begin{vmatrix} 1 & 0 \\ \sigma & 1 \end{vmatrix}.$$

Like Psheborski we want to name the class of polynomials satisfying the above-mentioned conditions as Z.

He began the way to the solution with the proposition that a solution of this problem *exists*.

5.1.4.1 Existence of a Solution

Psheborski showed that it suffices to regard only polynomials $\in Z$ with a priori bounded coefficients. Then the existence of a solution follows with Weierstraß' theorem because of the fact that the error of approximation is a continuous function of the coefficients.

We see,[19] that we can decompose the polynomials $p \in Z$ into summands $p =: h_1 + h_2$, where

$$h_1(x) := p_i x^i + p_k x^k, \quad (i \neq k)$$

with indices k and i suitably chosen for the additional condition and the rest h_2. Both linear equations can be decomposed into[20]

$$p_i = \frac{\alpha \beta_k - \beta \alpha_k}{\alpha_i \beta_k - \beta_i \alpha_k} \quad p_k = \frac{\alpha_i \beta - \beta_i \alpha}{\alpha_i \beta_k - \beta_i \alpha_k} \tag{5.1}$$

and

$$0 = \sum_{j=0 j \neq i j \neq k}^{n} p_j x^j.$$

Since zero is that function of type h_2 with minimal norm, it suffices to regard only functions of type h_1. But the equations (5.1) show that the norm of all these functions is bounded by $(\max\{|a|, |b|\}) \times (\max\{|p_i|, |p_k|\})$.

So there exists a minimal solution of the problem. We also recognized the rôle of the additional conditions.

---

[18] Sometimes the opposite is said. In fact Psheborski's results hold without this additional assumption, but Psheborski himself did not show this!

[19] Here we are a bit more detailed than Psheborski.

[20] The fractions are valid because of the additional conditions.

Obviously Psheborski did not know Kirchberger's thesis exactly.[21] In his proof of the continuity of the operator of best approximation it had already been proved[22] that a given bound for the norm of a polynomial already implies a bound for the coefficients of the polynomials, without any additional conditions.

5.1.4.2 Criteria for the Solution

So there exists a minimal solution $h \in Z$. Their deviation points we want to name as $x_1 < \cdots < x_p$, $p \leq n + 1$.

Then there followed theorems characterizing the minimal solution. They are similar to those of V. A. Markov:

Theorem 5.1 $h \in Z$ *is a minimal solution if and only if there does not exist a polynomial* $g \in \mathbb{P}_n$ *with*

1. $\omega(g) = \omega_1(g) = 0$ *and*
2. the numbers

$$\tau_i = h(x_i)g(x_i)$$

have an equal sign for all $i = 1, \ldots, p$.

This theorem was similarly proved like the respective theorem of V. A. Markov.

Pheborski also proved generalized alternation theorems which are similar to Theorem 3.9 on page 108 and use Lagrange's basic polynomials.

Psheborski's contribution was especially interesting because of the fact that he discussed the question of the existence of a solution in a detailed way. Indeed, this question was not unnecessary, since he was not able to calculate a minimal solution. V. A. Markov ignored this question, but his main result, the estimation for the kth derivative of a polynomial, holds independently from it.

Thus, Psheborski's approach was more modern, more 'European,' but nevertheless he was not able to prove a result which could be compared with that of V. A. Markov.

He presented his results in the Comptes Rendus [Psh13/3], where he also remembered the contribution of V. A. Markov, which was hardly available at that time. Another application of his results Psheborski published in [Psh14] for the special side-conditions

---

[21] So presumably Psheborski's interest in questions from approximation theory arose only after his stay in Göttingen. The stimulus might rather be V. A. Markov's work (as Buts has already assumed [But92, p. 123]), since it was re-discovered by S. N. Bernstein in 1912, who also was engaged in the publication of this work in the 'Mathematische Annalen' and introduced it with a comment [MarV16]. Compare also the following explanations to Bernstein's doctoral thesis [Bern12/2]).

[22] Compare section 4.3.1.1 on page 137.

$$p^{(k)}(z) = \sigma, \quad p^{(i)}(z) = \tau \quad \text{for fixed } z \in \mathbb{R}.$$

It seems that according to approximation theory Psheborski's main effort was the support of the scientific career of Sergey Natanovich Bernstein. He was the main supervisor of his Russian theses, the master thesis [Bern08] and the doctoral thesis [Bern12/2].

5.2 A Short Biography of Sergey Natanovich Bernstein

Sergey Natanovich Bernstein was born February 22, (March 5) 1880 in Odessa in the family of the physician Natan Bernstein, who was an extraordinary professor of the Novorossisk university.[23]

After grammar school Sergey Natanovich and his elder sister left Odessa at the request of their mother[24] to study in Paris. The sister would become a well-known biologist, stayed in Paris and worked at the Pasteur institute.

In 1899 Sergey Natanovich finished his studies of mathematics, but rejected this profession and decided to begin engineering studies at the Paris electrotechnical school. After his studies there he nevertheless continued to deal with mathematics and registered for the winter semester 1902/03 in Göttingen. There he mainly studied under the supervision of David Hilbert. He stayed three terms, returned to Paris in spring 1904 and defended there his thesis for the receipt of the docteur-es-sciences 'Sur la nature analytique des solutions des équations aux dérivées partielles du second ordre' [Bern04/1], where he gave an important contribution to the solution of the 19th Hilbert problem.[25] The report about the thesis was written by Picard.[26] Afterwards he participated in the international congress of mathematicians in Heidelberg[27] and stayed in Heidelberg for a while,[28] before he returned to Russia (St. Petersburg) in 1905.

The results of his thesis were published very early ([Bern03]—partial results—and [Bern04/2]), and so Bernstein became well known very fast. But

---

[23] The biographical data were mainly taken from [Bog91]. Some additions were made with the help of [Akh55] and sources from Göttingen archives ([Amt10] and [BernHil]). Sometimes the secondary sources are contradictory. In such a case we cited the work of Bogolyubov as a main source. The publication of an extensive biography of Sergey Natanovich Bernstein is planned at 'Nauka' by the authors A. N. Bogolyubov and O. N. Buts-Bondar so most of the indistinctions should be clarified at some future time.

[24] It is not clear when they moved to France.

[25] For an exact analysis of Hilbert's problems concerning Bernstein and Bernstein's contributions to them compare A. G. Sigalov's article in [Ale98, p. 259–274].

[26] Compare [Pic04].

[27] See [Ver04].

[28] At least until January 1905, as follows from the address Bernstein used in the first two letters to David Hilbert [BernHil, Nr. 1-2]. (Compare [Stef99]).

this was not of advantage to him in Russia: Foreign grades were not recognized there, and he had to pass the master exams.[29]

He passed them with difficulties, especially the exam at Korkin about differential equations caused trouble because he was forced to solve differential equations by classical means.[30]

Nevertheless in 1907 he could work as a professor at the recently founded polytechnical college for women because there formal requirements were not so important. Probably because of the non-satisfactory scientific atmosphere for him he decided to move to Kharkiv where in 1908 he defended his master thesis 'Investigation and Solution of Elliptic Partial Differential Equations of Second Degree.' His opponents were D. M. Sintsov[31] and A. P. Psheborski. This thesis included both a detailed elaboration of his approach to solve Hilbert's 19th problem and a contribution to Hilbert's 20th problem.

In Kharkiv he worked as a privat-dotsent and as a professor at the women's college. Nevertheless he was not content with his—as he said—'hopeless' situation.[32] He clearly intended to work abroad and obviously he got an offer from W. F. Osgood who was professor at Harvard University,[33] presumably which is why he went again to Göttingen in spring 1910. Probably he wanted to meet Jackson, who just then studied there, and to discuss this question with him. But these plans failed for an unknown reason. Bernstein returned to Kharkiv and stayed there.

In 1913 he defended his doctoral thesis which he had finished one year before 'About the Best Approximation of Continuous Functions by Polynomials of Given Degree.' The report was written by Psheborski [Psh13/1]. Already in 1911 he was awarded the prize of the Belgian academy of sciences for his results on this subject.

Until 1918 he gave lectures at the women's college and from 1912 until 1918 he additionally was a lecturer at the Kharkiv commercial university («коммерческий университет»). In 1920 he became ordinary professor at Kharkiv university.

Already in 1910 he had begun to deal with questions of didactics of mathemtics, also together with Sintsov.

---

[29] Compare the explanations in appendix B.1—the master thesis was more or less the same as the docteur-es-sciences.

[30] Compare [Ozhi68, Chapter 5]. There are rumours that he only passed the second time.

[31] Dmitri Matveevich Sintsov (1867–1946), studium in Kazan, 1899–1903 prof. at the higher mining institute of Ekaterinoslav (today Dnipropetrivsk/Dnepropetrovsk), since 1903 ord. prof. at Kharkiv university, since 1906 president of the Kharkiv Mathematical Society, since 1939 member of the Ukrainian academy of sciences. He worked on differential equations and differential geometry.

[32] Compare the fourth letter to D. Hilbert [BernHil, Nr. 4].

[33] Compare the sixth letter to D. Hilbert ([BernHil, Nr. 6]).

In the 1920s the system of scientific institutes and universities was reorganised[34], the institutes were divided into several chairs and the scientific was separated from the pedagogical work. One of the authors of these ideas had been Bernstein. At that time he was scientifically active mainly on two subjects: approximation theory ('Constructive Function Theory') and probability theory. Here he was the first to develop an axiomatic basis.

In 1922–24 he visited Germany and France and at the Sorbonne he gave two semesters of lectures about probability theory.

In 1924 Bernstein was elected corresponding member of the Russian Academy of Sciences, in 1925 ordinary member of the Ukrainian Academy. In 1928 he became director of the Kharkiv mathematical institute. In the same year he became corresponding member of the Paris Academy of Sciences, as a successor of Gösta Mittag-Leffler who died shortly before.

In 1929 he was elected ordinary member of the Academy of Sciences of the USSR, together with the mathematicians Nikolay Mitrofanovich Krylov,[35] Ivan Matveevich Vinogradov and Nikolay Nikolaevich Luzin. Simultaneously Dmitri Fedorovich Egorov and Dmitri Aleksandrovich Grave were elected honorary members of the academy. It is possible that this election protected him from prosecution in the following years.

In 1930 the dark time of the 'Leningrad Mathematical Front' began. Triggered by the polemic document [AndL31] certain groups began to purge mathematics of 'idealistic elements,' where they started with the Leningrad Mathematical Society.[36]

During the first All Union Congress of Mathematicians in Kharkiv (June 24–29, 1930) S. N. Bernstein was not spared a political interference into mathematics. A problem formulated for the congress was read as 'Application of the method of dialectic and historical materialism to the history of mathematics and its underpinning' and it was also suggested to apply this method to mathematical research. Bernstein wrote an answering letter in the journal of the Kharkiv physico-chemico-mathematical institute, which was spread in a large number of issues, where he proclaimed that dialectic materialism leads to mathematical illiteracy.[37] Consequently he was removed from his post as director of the mathematical institute, but remained there as a professor until 1932 in spite of all attacks against him.

---

[34] Compare also section 5.1.

[35] He should not be taken for Lyapunov's pupil Aleksey Nikolaevich Krylov, who had already been elected member of the academy in 1916.

[36] Naturally only after the end of the era of stagnation could a first try be made to reappraise the 1930s in a scientifically satisfactory manner. This cannot be a subject of the present work, so we only want to mention here Bogolyubov's short biography of Bernstein [Bog91] and the more general contributions [Erm98/2], [Pau97] and [Delo99].

[37] Cited after [Bog91, p. 60]: «диалектический материализм приводит к математической безграмотности».

In 1932 Bernstein left Kharkiv and became head of the department of probability theory and mathematical statistics of the mathematical institute of the academy of sciences which was in Leningrad at that time. From 1934 he also taught at the Leningrad university.

As Bogolyubov judges he left Kharkiv in time, for shortly after his leave the purge of Kharkiv university began. Many lecturers and students were arrested, some of them were shot.

At that time in Leningrad among others there studied and taught (also with Bernstein) the Grand Seigneur of modern approximation theory, G. G. Lorentz,[38] who also suffered from prosecution and was forced to leave his home country. Therefore, however, he could spread Bernstein's ideas in Western countries and help to overcome the isolation of Russian mathematics in the Soviet time.

We should also mention that Bernstein could also hardly be intimidated in the following years. In 1936, the culminating point of Stalin's purges at the affair 'Luzin' was tried,[39] Bernstein decisively defended Luzin's position. This was perilous at that time. On January 1, 1939 Bernstein started his lecturer job at Moscow university and was removed from being head of the mathematical institute, but he kept his home in Leningrad. In 1940 he became honorary member of the Moscow Mathematical Society.

The days of WWII he spent in the Kazakh town of Borovoe. His son German Sergeevich remained in Leningrad and died when he wanted to leave the town after the blockade.

His son's death was the stimulus for Bernstein to leave Leningrad forever, and he also moved his place of living to Moscow. In 1947 he was dismissed from the university and became head of the department of constructive function theory at the Steklov institute. He held this position until 1957.

---

[38] Georg(e) Günter (until 1946 Georgi Rudolfovich) Lorentz (*1910 St Petersburg), 1922–1926 school education in Tiflis, 1926–1928 student at the technological institute in Tiflis, 1928–1931 at mathematical-mechanical faculty of Leningrade university, diploma 1931, 1931–1936 teaching assistant univ. Leningrad, 1936 candidate thesis, 1936-1941 lecturer univ. Leningrad, 1941–1942 military service, 1942 escape to Kislovodsk in the Caucasus mountains, 1943 escape to Poland, 1944 invitation to Tübingen, there assistant's job, doctoral thesis and 'Habilitation,' 1946-1948 lecturer univ. Frankfurt/Main, 1948–1949 'Honorarprof.' univ. Tübingen, 1949 Research scholarship univ. Toronto, there instructor, later assistant prof., 1953–1958 full prof. Wayne State univ. Detroit, 1958–1969 full prof. univ. Syracuse (USA), 1968–1980 full prof. univ. of Texas in Austin, 1980 emeritus, 1997 edition of selected works by his son Rudolph A. Lorentz [Lor97] (which also contains an autobiography, where these data were taken from). George G. Lorentz lives in California.

[39] To say it in a few words, Luzin was accused of 'idealism' because his theories were alleged far from applications and was nearly arrested. Compare [Delo99]. A detailed comment can also be found in [Pau97].

His work was highly recognized abroad, especially in France: in 1944 he became honorary doctor of Algiers University,[40] in 1945 even of the Sorbonne, and in 1955 Bernstein was elected foreign member of the French Academy of Sciences.

In spite of his critical attitude to the ideals of Marixism-Leninism he was also highly decorated by the Soviet Union: Sergey Natanovich Bernstein was awarded the Lenin medal twice, once he got the medal of the 'merited red standard' («трудового красного знамени») and some of his contributions were awarded the Stalin prize. A special honour was the edition of his collected work in his lifetime.

Sergey Natanovich Bernstein died October 26, 1968 from after effects of an operation.

5.3 First Contributions to Approximation Theory

His biography, his first mathematical papers, but also his warm relation to Hilbert justify naming Sergey Natanovich Bernstein a pupil of David Hilbert, which does not contradict the fact that Hilbert was not Bernstein's official supervisor.[41]

Bernstein was a modern mathematician, not caught in classical thoughts like some representatives of the St Petersburg Mathematical School. His interest in questions from approximation theory had therefore firstly not been influenced by their ideas. Besides, his mathematical education had completely taken place abroad, and at first his talent had not adequately been acknowledged in St Petersburg.

And so it was one of the ironies of fate that with Bernstein's work, approximation theory again came to life in Russia.

5.3.1 A Proof of Weierstrass' Theorem

Already very early Bernstein had been interested in finding a simple proof for Weierstraß' approximation theorem, and he had also been one of those mathematicians who tried to prove the convergence of the Lagrange interpolation algorithm, as he himself admitted much later, at the 1930 congress of mathematicians in Kharkiv:[42]

---

[40] During the German occupation Algiers University replaced the universities of the independent French republic.

[41] Why Bernstein did not defend his first (French) doctoral thesis in Germany, is not known. It might have simply been connected with the fact that his German was not good enough—his statements about his knowledge of languages in the fourth letter to D. Hilbert [BernHil, Nr. 4] might be an indication for that.

[42] His main speech was printed two years later as [Bern32] in the 'Communications of the Kharkiv Mathematical Society and the Ukrainian Mathematical Institute'.

"I remember that one of my first mathematical investigations was to try to prove this [Weierstraß approximation theorem, K.G.S.] starting from Lagrange's formulae. But, as we now know, this attempt could not have been successful because Lagrange's formula in general diverges."[43]

He did not say exactly when he made this attempt, but presumably it was before 1904, when in his talk during the Heidelberg congress Borel presented an example of a non-convergent sequence of Lagrange polynomials—obviously without any knowledge of Runge's work [Run01] from 1901. Bernstein participated in this congress and so surely got to know Borel's result. Maybe this problem generally stimulated him to deal with approximation theory.

In 1916 he returned to the question of convergence of Lagrange's algorithm and proved [Bern16]:

Theorem 5.2 *If n is an arbitrary natural number and $R \in \mathbb{P}_n$ Lagrange's interpolation polynomial of degree n interpolating in the nodes of the Chebyshev polynomial T_n, then for any continuous function $f : [-1,1] \to \mathbb{R}$ there exists a constant $K \in \mathbb{R}$ so that there holds:*

$$\|f - R\|_\infty \leq K E_n(f) \log n.$$

If f even satisfies the Dini–Lipschitz condition, then with Lebesgue's result [44] it is clear that Lagrange's procedure converges with these assumptions.

Essentially more well known is Bernstein's second and successful attempt from 1912 to prove Weierstraß' approximation theorem [Bern12/1], where also the now so-called 'Bernstein polynomials' had been introduced.

For a given continuous function $f : [0,1] \to \mathbb{R}$ they are defined by

$$B_n(f) := \sum_{i=0}^{n} f\left(\frac{i}{n}\right) \binom{n}{i} x^i (1-x)^{n-i}. \tag{5.2}$$

With Bernstein then there holds:

Theorem 5.3 *For all $f \in C[0,1]$ we have:*

$$\lim_{n\to\infty} \|f - B_n(f)\|_\infty = 0.$$

The representation of these polynomials shows a connection between these polynomials and the expected value of Bernoulli's distribution. And so Bernstein proved this theorem using arguments from probability theory. It is very difficult to summarize the proof, therefore we want to give it completely:

---

[43] [Bern32, p. 21]: «Згадую, що в одній з моїх перших наукових спроб я хотів довести це, виходячи з інтерполяційної формули Lagrange'a. Отже, такий процес, як це ємо тепер, не міг мати успіху, бо Lagrange'ева формула взагалі є розбіжна.»

[44] Compare Table 4.7.

Proof Let A be an event which happens with probability x. If a player is paid the amount of $f\left(\frac{i}{n}\right)$, when A comes i times in n attempts, then the expected profit of the player is with Bernoulli just $B_n(f)$.

Let $\varepsilon > 0$ be an arbitrary, but fixed number. Because of the continuity of f (and the compactness of $[0,1]$) theorem exists a number δ so that for all $x, y \in [0,1]$ there holds:

$$|x - y| < \delta \Rightarrow |f(x) - f(y)| < \frac{\varepsilon}{2}.$$

Let η be the probability that a random number $x \in [0,1]$ does not fall in one of the intervals $[\frac{i}{n} - \delta, \frac{i}{n} + \delta]$, so

$$\eta := P\left(\forall i \in \{1, \ldots, n\} : |x - \frac{i}{n}| > \delta\right).$$

Now let $x \in [0,1]$ be arbitrarily chosen. Let us assume that x lies in one of the intervals $[\frac{i}{n} - \delta, \frac{i}{n} + \delta]$, let's say in $I(x)$. The probability for that is $1 - \eta$. Let

$$U(x) := \min_{y \in I(x)} f(y), \quad V(x) := \max_{y \in I(x)} f(y).$$

Then we get the inequalities

$$U(x)(1 - \eta) - \eta\|f\|_\infty < B_n(f) < V(x)(1 - \eta) + \eta\|f\|_\infty,$$

since the maximal loss is $-\|f\|_\infty$ and the maximal profit is $\|f\|_\infty$. They can also be written as:

$$f(x) + \underbrace{U(x) - f(x)}_{<\frac{\varepsilon}{2}} - \eta\left(\|f\|_\infty + U(x)\right) < B_n(f) \quad \text{and}$$

$$f(x) + \underbrace{V(x) - f(x)}_{<\frac{\varepsilon}{2}} - \eta\left(\|f\|_\infty + V(x)\right) > B_n(f). \tag{5.3}$$

With the large number theorem we choose n as sufficiently large so there will hold:

$$\eta < \frac{\varepsilon}{4\|f\|_\infty},$$

and so (5.3) turns out to be

$$f(x) - \varepsilon < B_n(f) < f(x) + \varepsilon.$$

Since these inequalities hold for all $x \in [0,1]$, the theorem has been proved.

5.3.2 A Prize Competition of the Belgian Academy of Sciences

We have already mentioned that in Western Europe an interest in getting quantified results of Weierstraß' approximation theorem arose.

To support this aim the Belgian Academy of Sciences on a suggestion by de la Vallée-Poussin, had posed the following problem in 1903 as a competition:

"Present investigations into the development of real or analytical functions into power series."

Then de la Vallée-Poussin had specified, as well:

"Is it possible or not to approximate the ordinate of a polygonal line, or—what is the same—the function $|x|$ on the interval $[-1, 1]$ by a polynomial of nth degree with a higher degree of approximation than $\frac{1}{n}$?"[45]

That the approximation of $|x|$ played an important rôle in the approximation of continuous functions was clear at least since Lebesgue's proof of Weierstraß' approximation theorem [Leb98], for he proved it firstly showing that an arbitrary function can be approximated by polygonal lines and then approximating the lines by polynomials.[46]

Only after seven years did de la Vallée-Poussin show a first result [Val10]. He proved that there holds:

$$E_n(|x|) > \frac{k}{n \log^3 n}.$$

After another year Sergey Natanovich Bernstein gave a first indication that this question has to be answered with 'no' with the following theorem [Bern11]:

Theorem 5.4 *There exists a number k so that there does not hold for any natural number n :*

$$E_n(|x|) < \frac{k}{n \log n}.$$

He could give a complete answer in [Bern12/4] which then was printed by the Belgian academy as the prize-winning treatise.

There Bernstein could give an exact order of convergence of $E_n(|x|)$. Since $|x|$ is an even function, its best approximation must be an even function, too, and so the following theorem suffices for the exact order:

---

[45] We had no access to the Belgian original document and cited after [Gon45, p. 158]: ,,Представить новые исследования, касающиеся разложения функций действительных или аналитических в ряды полиномов. [...] Возможно или нет представить ординату полигональной линии, или, что сводится к тому же, $|x|$ в промежутке $[-1, 1]$ посредством полинома степени n с приближением более высокого порядка, чем $\frac{1}{n}$?"

[46] To be precise, this idea was due to Runge, who used it to show the approximation theorem named after him [Run85/2].

Theorem 5.5 *There exists a natural number n_0 so that for arbitrary $n \in \mathbb{N}$ there holds:*

$$\frac{0.278}{2n} < E_{2n}(|x|) < \frac{0.286}{2n}.$$

But his papers on this subject reached farther. The contributions [Bern11], [Bern12/2], [Bern12/3], [Bern12/4] and [Bern13/2] are partly devoted to questions of this kind, [Bern12/3] and [Bern13/2] even exclusively.

This made clear that Bernstein regarded this subject as of similar importance as the Belgians had done with their posing of the competition. He stated in 1912 in a speech to the international congress in Cambridge (UK):

> "The example of the problem of the best approximation of the function $|x|$, suggested by de la Vallée-Poussin, again confirmed the fact that a well-posed single question is able to be the starting point for far reaching theories"[47]

Also Bernstein's first article, where he discussed questions of approximation theory, had the character of a forerunner for these questions. It is the contribution ,Sur l'interpolation' [Bern05] published in 1905.

Here he dealt with the question how well a continuous function could be approximated by interpolating polygonal lines. In those days the question of the convergence of interpolating algorithms had not been answered.

Replying to Runge's result according to the convergence of the Lagrange algorithm in [Run01], he mentioned that the practical man often wants to approximate a function by polygonal lines. So regularity assumptions did not always correspond with the interest of practice.

Bernstein was able to show that such an interpolation process uniformly converges for twice continuously differentiable functions. He used the representation of the interpolant by means of multiple differences.

5.3.3 The Prize-Winning Treatise

The above-cited words about the relevance of the problem of best approximation of $|x|$ had a concrete background: As the prize-winning treatise Bernstein not only presented a solution of the posed problem, but the extensive monograph ,Sur l'ordre de la meilleure approximation des fonctions continues par des polynomes de degré donné' [Bern12/4], which reached much farther and counted as the foundation of the later so-called 'Constructive Function Theory'. He submitted it in June 1911;[48] after one year the Russian translation

---

[47] [Bern12/2, Russian translation, p. 117]: «Пример задачи о наилучшем приближении $|x|$, предложеноой Валле Пуссеном, дает еще одно подтверждение того факта, что хорошо поставленный вопрос способен быть отправной точкой для далеко идущих теорий.»

[48] Bernstein said this in a footnote to his doctoral dissertation [Bern12/2]. He wrote: "Except for two additions to the chapter IV and V the present work is a trans-

was also printed. This treatise served him as a doctoral dissertation, which he defended on May 19th, 1913.

The fact that the Russian translation played the rôle of a second edition might have caused it to appear worse than the French version. At least Psheborski had this opinion:

"It seems that the enormous engagement in the publication of his French edition was the reason for the large carelessness of the present work which often has the character of a manuscript written 'for himself'."[49]

After some inequalities which have the same shape as those of A. Markov and V. Markov and the explanations about the approximation of $|x|$, Bernstein proved here a series of inverse theorems which had indeed had the fundamental character of a new view on the theory of functions. They showed that differentiability and analyticity properties may follow from the speed of approximation of a function.

To be precise, we find there the following results for real-valued functions $f \in C[a, b]$:

Theorem 5.6 *If there is*

$$\lim_{n \to \infty} E_n(f) \log n = 0,$$

then f satisfies the Dini–Lipschitz condition.

Using Lebesgue's result from [Leb10/1] we have then the characterization

Corollary 5.7 (Bernstein 1911–Lebesgue 1910)

f satisfies the Dini–Lipschitz condition

$$\Longleftrightarrow$$

$$\lim_{n \to \infty} E_n(f) \log n = 0.$$

The next two results are the inverse versions of two theorems due to Jackson.

lation of the work with the same title which has been awarded the prize of the Belgian Academy of Sciences, where I had submitted it in June 1911." ([Bern52, p. 12] «Настоящая моя работа, за исключением двух ,,Добавлений'' к IV и V главам, представляет, с незначительными редакционными изменениями, перевод мемуара под тем же заглавием, удостоенного премии Бельгийской академии, куда он был направлен мною в июне 1911 г.»)

[49] [Psh13/1, p. 27]: «Быть может, то обстоятельство, что автор особенно тщательно заботился о редакции своей французской работы, явилось причиной большой небрежности в редакции рассматриваемого сочинения, носящего подчас характер черновых заметок ,,для себя''.»

Theorem 5.8

1. *If the series*

$$\sum_{n=1}^{\infty} E_n(f)n^{p-1}$$

converges, then f is p times continuously differentiable.

2. *If for an $\alpha > 0$ the series*

$$\sum_{n=1}^{\infty} E_n(f)n^{p-1+\alpha}$$

converges, then $f^{(p)} \in \text{Lip}_\alpha$.

Both assumptions are stronger than Jackson's necessary conditions, so we don't have characterizing theorems, but only:

Corollary 5.9 (Bernstein–Jackson 1911)

1. *There holds:*

$$\sum_{n=1}^{\infty} E_n(f)n^{p-1} < \infty$$
$$\Rightarrow f \in C^p[a,b]$$
$$\Rightarrow \lim_{n \to \infty} E_n(f)n^p = 0.$$

2. *Let $\alpha > 0$. Then:*

$$\sum_{n=1}^{\infty} E_n(f)n^{p-1+\alpha} < \infty$$
$$\Rightarrow f^{(p)} \in \text{Lip}_\alpha[a,b]$$
$$\Rightarrow \lim_{n \to \infty} E_n(f)n^{p+\alpha} = 0.$$

The last theorems via $n \to \infty$ led to theorems characterizing infinitely often differentiable and real-analytic functions which were proved by Bernstein alone:

Theorem 5.10 (Bernstein 1911)

1. *f is infinitely many times differentiable if and only if for all $p \in \mathbb{N}$ there holds:*

$$\lim_{n \to \infty} E_n(f)n^p = 0.$$

2. *f is a real-analytic function, if and only if there exists a number $\rho > 1$ with:*

$$\lim_{n \to \infty} E_n(f)\rho^n = 0.$$

With these new results approximation theory showed a new character. If it was called by Chebyshev and his pupils 'theory of functions deviating the least possible from zero' and its aim was only the investigation of extremal properties of certain polynomials, now it had a new meaning in the theory of real-valued functions as a whole.

The concepts of Lipschitz continuity, (continuous) differentiability and analyticity, independently introduced, could now be described by the speed of the operator of best approximation, and partially even characterized.

5.3.4 A Brief Note about the Interrelation between Jackson's and Bernstein's Contributions

We have already established that Bernstein's main early contributions to approximation theory, [Bern11] and [Bern12/4], were finished nearly simultaneously with Jackson's thesis [Jack11]. The results of all of these papers were very similar, but both authors stated that they found their results independently.[50]

The fact that they both stayed in Göttingen during the summer semester 1910, at a time when both had already begun their investigations on the same subject, makes these statements not very credible. We remember that there was an objective reason for Bernstein to meet Jackson, namely the question for a possible lecturer's job in Harvard. Therefore they probably met. Did they talk about their research on quantitative approximation theory? Of course possibly they did not.

---

[50] Jackson wrote [Jack11, p. 12:] "A few weeks before the submission date of the prize-winning treatise S. Bernstein published a series of theorems, which are partially similar to some of the theorems, which will follow [...] Besides this it contains theorems, which do not occur in this treatise." („Wenige Wochen vor dem Einlieferungstermin der Preisarbeit hat S. Bernstein in einer kurzen Note ohne Einzelheiten seiner Beweise eine Reihe von Stzen verffentlicht, die zum Teil einigen derer, die hier folgen sollen, sehr hnlich sind. [...] Auerdem erhlt sie auch Stze, die in dieser Abhandlung gar nicht vorkommen."). Bernstein wrote in the speech defending his doctoral thesis about one of these very similar results (here the determination of the exact constant for the approximation of $|x|$) [Bern12/2]: "the same result was reached (independently from me) a little later by Jackson" («тот же результат (независимо от меня) немного позднее получен Джэксоном.»

5.3.5 Quasianalytic Functions

Unfortunately the gaps between the two statements of Theorem 5.9 cannot be closed without weakening the assumptions,[51] because counterexamples exist for both directions, as Bernstein himself mentioned [Bern13/1].

This was the reason why the question of the speed of convergence of $|x|$ was so difficult to answer. After all there holds

$$\frac{0.278}{n} < E_n(|x|) < \frac{0.286}{n},$$

and so the series

$$\sum_{n=0}^{\infty} E_n(|x|)$$

must diverge, although $|x| \in Lip_1[a, b]$.

For a better handling with these gaps Bernstein introduced the concept of *quasianalytical* functions [Bern14]:

Definition 5.11 *A function $f \in C[a, b]$ is called quasianalytical, if there exist a number $\rho < 1$ and a sequence $(\alpha_n)_{n \in \mathbb{N}}$ so that $\forall n \in \mathbb{N}$:*

1. $\alpha_n \leq \rho^n$
2. $E_{\alpha_n}(f) \leq \alpha_n$.

In fact there are non-differentiable quasianalytic functions, for instance $\phi : [-1, 1] \to \mathbb{R}$ with:

$$\phi(x) := \sum_{n=0}^{\infty} \frac{\cos(F(n) \arccos x)}{F(n)},$$

where is set

$$F(0) := 1 \quad F(n+1) := 2^{F(n)}.$$

Since its summands are Chebyshev polynomials it is well defined, and secondly there holds

$$E_{F(k)}(\phi) \leq \|\phi - F(k) T_{F(k)}\|_\infty = \| \sum_{n=k+1}^{\infty} \frac{1}{F(n)} T_{F(n)} \|_\infty$$

$$\leq \frac{1}{F(k)} \sum_{n=k+1}^{\infty} \frac{1}{2^{n+1}} < \frac{1}{2^k},$$

---

[51] In 1919 de la Vallée-Poussin (not Bernstein!) could prove a necessary and sufficient condition for 2π-periodical, α-Lipschitz continuous functions [Val19, p. 57f.], where he regarded the trigonometric case and had to exclude $\alpha = 1$. His theorem then read as follows:

$$f^{(p)} \in Lip_\alpha \Leftrightarrow \lim_{n \to \infty} E_n(f) n^{p-1+\alpha} = 0.$$

thus, ϕ is quasianalytical. Differentiating each summand we get for $x = 0$ the divergent series

$$1 + \sum_{n=1}^{\infty} \sin\left(F(n)\arccos 0\right) = 1 + \sum_{n=1}^{\infty} \sin\frac{\pi}{2}F(n),$$

and so ϕ is not differentiable.

Another possibility to define quasianalytical functions came from the definition of higher moduli of continuity, which Bernstein introduced[52] in [Bern12/4]. They were defined by

$$\omega_2(f,\delta) := \max_{|h| \leq \delta} |f(x + 2h) - 2f(x + h) + f(x)|,$$
$$\omega_3(f,\delta) := \max_{|h| \leq \delta} |f(x + 3h) - 3f(x + 2h) + 3f(x + h) - f(x)|,$$

$$\vdots \quad \vdots \quad \vdots$$

Then Bernstein defined

Definition 5.12 *A function $f \in C[a,b]$ satisfies the generalized Lipschitz condition of degree α and the order i, if there exist a constant K and a null sequence $(\varepsilon_k)_{k \in \mathbb{N}}$ so that*

$$\omega_i(f,\varepsilon_k) \leq K\varepsilon_k^{\alpha}$$

for all $k \in \mathbb{N}$.

Then an intersecting theorem is

Theorem 5.13 *If for a function $f \in C[a,b]$ there exist a constant K, a natural number p and a sequence $(n_k)_{k \in \mathbb{N}}$ so that for all $k \in \mathbb{N}$ there holds*

$$E_n(f) \leq \frac{K}{n_k^p},$$

then f satisfies the generalized Lipschitz condition of order i and degree

$$\alpha_i := \frac{ip}{i + p}$$

for all $i \in \mathbb{N}$.

---

[52] The definition of higher moduli of continuity is often wrongly attributed to Zygmund [Zyg45].

5.4 Constructive Function Theory as the Development of Chebyshev's Ideas

As we have already seen analysing several European contributions, questions of quantitative approximation theory could be regarded as a direct consequence of Weierstraß' theorem. Until Bernstein entered the scene all quantitative theorems had been derived without the help of the results of the St Petersburg Mathematical School. This was no longer true for Bernstein's inverse theorems.

Bernstein proved the sufficiency of the conditions of Theorem 5.9 via V. Markov's inequality, since the convergence of the series

$$\sum_{n=1}^{\infty} E_n(f) n^{p-1}$$

implies the possibility to represent it as an absolutely convergent polynomial series which can be differentiated in ranks as long as V. Markov's inequality allows. So this inequality became one of the most important means of Bernstein's work.

Inequalities of this kind were the intial point of Bernstein's lectures which he gave in 1926, calling them ‚Leçons sur les propriétés extremales et la meilleure approximation des fonctions analytiques d'une variable réelle'. He published them in the same year ([Bern26], Russian version and revision [Bern37]).

So the results of the St. Petersburg Mathematical School and those from Western European authors originated by Weierstraß were put together in Bernstein's work.

No wonder then that in his contribution [Bern32] Bernstein tried to prove Weierstraß' approximation theorem only by algebraic means.

There holds the following theorem:

Theorem 5.14 *Let $f \in C[a,b]$. If for a $n \in \mathbb{N}$ and a polynomial $p_n \in \mathbb{P}_n$ there holds the inequality*

$$\|f - p_n\|_\infty < 4L,$$

then there is an $n_0 > n$ and a polynomial $p_{n_0} \in \mathbb{P}_{n_0}$, for which there holds:

$$\|f - p_{n_0}\|_\infty < 3L.$$

This theorem can simply be proved by a shift of the difference $f - p_n$ with the help of a polynomial which in certain subintervals does not fall below the value L and does not exceed $2L$.

It is an argument of the type Chebyshev had often used before!

It is clear that this theorem is equivalent to Weierstraß' approximation theorem, since it guarantees the existence of a sequence $(\mu_n)_{n\in\mathbb{N}}$, for which there holds $\lim_n \mu_n = \infty$, and for which

$$\|f - p_{\mu_n}\|_\infty \le \left(\frac{3}{4}\right)^{\mu_n} L.$$

And thus, Sergey Natanovich Bernstein described the 'Constructive Function Theory' founded by him as the development of Chebyshev's ideas. The speech [Bern45/2] which the following part was taken from, even carried such a title:

"As constructive function theory we want to call the direction of function theory which follows the aim to give the simplest and most pleasant basis for the quantitative investigation and calculation both of empirical and of all other functions occurring as solutions of naturally posed problems of mathematical analysis (for instance, as solutions of differential or functional equations). In its spirit this direction is very near to the mathematical work of Chebyshev; therefore no wonder that modern constructive function theory uses and develops the ideas of our deceased famous member."[53] in a high extension[54]

The development of Chebyshev's theory was the application of analytical methods to its problems. In this way it could serve the mathematical analysis itself developing a deeper understanding of its concepts.

Whereas we partially rejected the comparison between Chebyshev and Euler in the summarizing section about Chebyshev's work because he was not interested in questions about the foundations of mathematics, we now state that in spite of his scepticism his theory also could enrich them.

From a purely applied problem there arose deep insights into pure mathematics; practice created a new theory.

---

[53] The speech was given before the academy whose ordinary member Chebyshev had been since 1859.

[54] [Bern45/2, p. 145]: «Конструктивной теорией функций мы называем направление теории функций, которое ставит себе целью дать возможно более простую и удобную основу для качественного изучения и вычисления как эмпирических функций, так и всяких функций, являющихся решениями естественно поставленных задач математического анализа (например, решений дифференциальных или функциональных уравнений). Это направление весьма близко по духу математическому творчеству Чебышева; не удивительно поэтому, что современная конструктивная теория функций в большой степени использует и развивает идеи нашего покойного сочлена.»

A

Biographies of Other Representatives of the Saint Petersburg Mathematical School

As we explained at the beginning of the third chapter, here we want to give some biographical data about the 'direct successors' of Chebyshev. Together with Korkin, Sochocki, Zolotarev, Posse, A. Markov sr. and V. Markov they formed the first generation of the St Petersburg Mathematical School.[1] Their subjects of interest were far from approximation theory, therefore we did not take them into consideration in the third chapter.[2]

A.1 Matvey Aleksandrovich Tikhomandritski

Matvey Aleksandrovich Tikhomandritski was born January 29, 1844 in Kyiv as a son of Aleksandr Nikitich Tikhomandritski who was professor at the chair of applied mathematics at the university of the holy Vladimir.[3] In 1848 his father accepted the position of the inspector of the St Petersburg pedagogical institute.

In 1861 Matvey Tikhomandritski registered at the physico-mathematical faculty of St Petersburg University; in 1865 he won a gold medal for a treatise about parabolic interpolation (least-square-approximation) in a students' competition. His paper was accepted then as a candidate thesis.

Form 1867 until 1879 he worked as a grammar school teacher; during this time in 1876 he defended his master thesis 'About hypergeometric series'

---

[1] Some of these representatives can be included in the second generation, since Chebyshev was their scientific 'grandfather' as they were pupils of his pupils. So we want to refer to our discussion of the name 'pupil' from the beginning of the third chapter. Biographical data about A. V. Bessel is missing because of the lack of sources about his person.

[2] We remember that this also holds for the mathematicians Posse and Sochocki. But these had a formative influence on the school itself: Posse as a teacher, Sochocki as a representative of function theory. Most of the persons mentioned here did not stay sufficiently long at St Petersburg to play a similar rôle.

[3] Sources: [Bio96], [Ozhi68, p. 125], [VorSPb] and [BoBu87, p. 501 f.].

under supervision of Korkin. From 1876 until 1879 he also taught at the institute of transportation. From 1879 until 1883 Tikhomandritski lectured as a privat-dotsent at the St Petersburg University (about elliptic functions and descriptive geometry). In 1883 he was appointed ordinary lecturer at Kharkiv University. There he was promoted to extraordinary professor in 1885 after the defense of the doctoral thesis 'Inversion of hyperelliptic integrals' and in 1888 he was promoted to ordinary professor. He lectured in Kharkiv until 1904.

The subjects of his lectures spread over nearly all subjects of education in pure mathematics of that time (for instance, 'higher algebra', 'calculus,' 'variational calculus,' ' theory' and 'functions of a complex variable'). His main subjects of research were higher algebra and elliptic functions and did not change for his whole life.

Matvey Aleksandrovich Tikhomandritski died in Kyiv in 1921.

A.2 Nikolay Yakovlevich Sonin

Nikolay Yakovlevich Sonin was born February 10 (22), 1849 in Tula[4]. Soon after his birth his family moved to Moscow because his father began the job of a lawyer there.

In 1865 Sonin registered at the physico-mathematical faculty of Moscow University. In 1869 he was awarded the gold medal for a work about 'Theory of functions of an imaginary variable'. This treatise was accepted as a candidate thesis. In 1871 he defended his master thesis 'About the Series Expansion of Functions'; in 1874 there followed his doctoral thesis 'About the Integration of Partial Differential Equations of Second Order,' which would be published in the 'Annalen' later in 1897. His main subject of interest was differential equations, especially his investigations about Bessel functions were paid great attention to.[5]

In 1871 Sonin began his lecturer's activities at the women's college, attributed to the third Moscow boy's grammar school; after one year he was appointed lecturer in Warsaw. In 1877 he was promoted extraordinary, in 1879 ordinary professor of Warsaw University. He mainly taught mathematical physics.

In 1891 he became emeritus after a twenty year lecturer job, but continued to give lectures.

In the same year he was elected corresponding member of the academy of sciences; by the initiative of Chebyshev he was elected ordinary member in 1893. According to the memories of D. A. Grave (comp. [Dob68, p. 11]) with this initiative Chebyshev wanted to prevent the election of Korkin.

---

[4] Sources: [Pos15] and [Son54].
[5] An overview of Sonin's work on Bessel functions can be found in [Son54].

In 1894 Sonin moved to St Petersburg and gave lectures at the women's university and at St Petersburg University (as privat-dotsent) about partial differential equations.

In 1899 he edited the collected works of P. L. Chebyshev [Chebgw1] together with A. Markov.

Besides these activities Sonin was engaged in science policy. Several times he was president of research commissions of the universities; in 1901 he became president of the research committee of the ministry of education.

Nikolay Yakovlevich Sonin died February 14 (27), 1915 in St Petersburg.

A.3 Aleksandr Vasilevich Vasilev

Aleksandr Vasilevich Vasilev was born July 2, 1853 in Kazan.[6] In 1870 he began his studies at the physico-mathematical faculty of St Petersburg University, which he finished in 1874 with a candidate thesis which was awarded with a gold medal.

Afterwards he left St Petersburg and returned to his home town Kazan. There he became privat-dotsent in 1875 and since 1887 he gave lectures as an ordinary, from 1899 as a merited professor.

In 1879 he made a trip abroad to Berlin and Paris, where among others he met Weierstraß, Kronecker and Hermite.

Vasilev was one of the founders of the Kazan Mathematical-Physical Society. He was its president from its foundation in 1890 until 1905.

In 1907 he returned to St Petersburg and gave lectures at the women's university. After the revolution in 1923 he became ordinary professor at Moscow State University.

His mathematical work spread over the subjects of algebra and potential theory, but Vasilev became famous rather by his work about the history of mathematics: he was the editor of the collected works of Lobachevski about geometry and the first biographer of Pafnuti Lvovich Chebyshev [Vas00].

Aleksandr Vasilevich Vasilev died October 9, 1929.[7]

A.4 Ivan Lvovich Ptashitski

Ivan Lvovich Ptashitski was born in 1854 in the province of Vilna (today Vilnius/Lithuania).[8]

---

[6] Sources: [Bazh88] (Biography of his son Nikolay Aleksandrovich Vasilev) and [BoBu87, p. 97]

[7] We could not find out, where Vasilev died. In Russian encyclopedic works there is often nothing said about the place of death. This remark also holds for some others persons mentioned here.

[8] Sources: [Bio96] und [Ozhi68].

In 1872 he registered at the physico-mathematical faculty of St Petersburg University. He defended his candidate thesis in 1876; in 1881 he defended his master thesis 'About the integration of irrational differentials in finite representation' under supervision of Korkin. His doctoral thesis he defended under supervision of Sochocki in 1888.

From 1880 until 1890 Ptashitski worked as a grammar school teacher in Peterhof; since 1882 he taught at the St Petersburg University as a privatdotsent. The subjects of his lectures were 'elliptical functions,' 'analytical geometry' and 'descriptive geometry.'

From 1890 he also lectured at the academy of ordnance.

His main subject of research was integration theory, especially elliptical functions.

Ivan Lvovich Ptashitski died in 1912.

A.5 Dmitri Fedorovich Selivanov

The only mathematician of the first generation of the St Petersburg Mathematical School who had been directly influenced by Weierstraß and his Berlin school was Dmitri Fedorovich Selivanov. But since he was only engaged in number theory and algebra, this had no influence on the development of approximation theory.

Dmitri Fedorovich Selivanov was born in 1855 in a small town in the province Penza.[9]

From 1873 until 1877 he studied mathematics at St Petersburg University and defended his candidate thesis 'About simply closed curves' in 1878. Then he stayed two years at the university to prepare for a lecturer's job. In 1880 he left the university for studies abroad. One year he spent in Paris at the chair of Hermite, afterwards he stayed two years in Germany to study function theory and higher algebra under supervision of Weierstraß and Kronecker.

In 1885 he defended his master thesis 'Theory of algebraic solutions of equations.' His doctoral thesis 'About Equations of fifth degree with integer coefficients' followed in 1889.

Since 1886 he lectured at St Petersburg University, at first as a privatdotsent; from 1889 he also gave lectures at the women's university, since 1891 at the technological institute. Only after 19 years of work as a privat-dotsent did he become extraordinary professor of the St Petersburg University in 1905.

Dmitri Fedorovich Selivanov died in 1932.

---

[9] Sources: [Andr90-05, vol. 29, p. 352], [Bio96] and [VorSPb]. The dates are often contradictory. In such a case we cited [Bio96], since this book is based on the personal information of the lecturers of St Petersburg University. The unreliable sources caused the fact that our biographical outline after 1891 consists only of dates from the university calendar [VorSPb].

A.6 Aleksandr Mikhaylovich Lyapunov

Surely one of the most prominent representatives of the St Petersburg Mathematical School was Aleksandr Mikhaylovich Lyapunov. His investigations about the stability of three-dimensional motions founded a new branch of the theory of differential equations. The approximation theory, however, did not play any rôle in his scientific work. Besides this his influence on the centre of the St Petersburg Mathematical School was restricted because he taught in Kharkiv for a long time, before he returned to St Petersburg as an academician.

Aleksandr Mikhaylovich Lyapunov was born May 25 (June 6), 1857 in Yaroslavl as a son of the head of the Yaroslavl observatory, Mikhail Vasilevich Lyapunov.[10]

In 1870, two years after the death of the father, Lyapunov's mother and her three sons moved to Nizhni-Novgorod, there in 1876 Aleksandr finished grammar school with a gold medal. In the same year he registered at the physico-mathematical faculty of St Petersburg University, at first at the department of natural sciences, in 1877 at the mathematical department. He was especially impressed by Chebyshev himself, both as a teacher and as a scientist. But his candidate dissertation he wrote under supervision of the physicist D. K. Bobylev. For this work he was awarded the gold medal in 1880 and could finish his studies. These investigations laid down the foundations of his first two publications 'About the balance of heavy bodies in heavy liquids which are located in a fixed vessel' and 'About the potential of hydrostatic pressure' (1881).

After his studies he remained at the university to prepare for a lecturer's job. In 1882 he finished the master examinations and started with the work which he won fame with: 'About the stability of ellipsoid balance states of a rotating liquid'. To this work he was stimulated by Chebyshev, but his official opponent at the defense of 1885 again was D. K. Bobylev. Already shortly after publication of the work a brief overview was published in the French 'Bulletin astronomique'. In 1904 the complete French translation ,Sur la stabilité des figures ellipsoïdales d'equilibre d'un liquide animé d'un mouvement de rotation' was published in the annals of Toulouse University.

In 1885 he was appointed privat-dotsent of St Petersburg University, but he moved to Kharkiv in the same year to take the chair of mechanics at the university there. In 1892 he defended his doctoral thesis 'The General Problem of Stability of Motion'. Its meaning even exceeded his master thesis, and so also this work was completely translated into French (,Problème général de la stabilité du mouvement' - Annales Toulouse 1907).

From 1899 until 1902 Lyapunov was president of the Kharkiv Mathematical Society; in 1900 he was elected corresponding member of the Russian academy of sciences; in 1901 he was elected ordinary member and took the

---

[10] Source: [Smi53].

position for applied mathematics as a successor of Chebyshev. This position
had been vacant since Chebyshev's death. Therefore he returned to St Petersburg and completely devoted himself to research. Some contributions to
stability theory would then follow.

Firstly during the revolutionary turmoil because of which he moved to the
Ukraine and Southern Russia, he again started to give lectures. In September 1918 he lectured at the university of Novorossiysk 'about the shape of
celestial bodies.' This lecture, however, was not finished. November 3, 1918,
three weeks after the death of his wife, Aleksandr Mikhaylovich Lyapunov
committed suicide.

A.7 Ivan Ivanovich Ivanov

Ivan Ivanovich Ivanov was born August 11, 1862 in St Petersburg.[11]

In 1886 he finished his studies of mathematics at St Petersburg University with his candidate thesis 'About prime numbers.' In 1891 there followed
his master thesis 'integral complex numbers' and in 1901 his doctoral thesis
'About some questions in connection with the number of prime numbers.'

Since 1891 Ivanov lectured at the St Petersburg University, since 1896 at
the women's university and since 1902 at the polytechnical institute.[12]

In 1924 Ivanov was elected corresponding member of the academy of sciences of the USSR.

After Georgi Feodosevich Voronoy, Ivanov counts as the most important
successor of Chebyshev on the subject of number theory.

Ivan Ivanovich Ivanov died December 17, 1939.

A.8 Dmitri Aleksandrovich Grave

Dmitri Aleksandrovich Grave counts as the founder of the Kyiv algebraic
school, which was the first algebraic school in the Soviet Union (by [Dob68]).
Among his pupils there were two of the most important Soviet representatives of approximation theory, Naum Ilich Akhiezer and Mark Grigorevich
Kreyn. Grave himself, however, did not write much about this subject. The
only exception was his solution of a problem from cartography proposed by
Chebyshev in 1856, which we have already mentioned in section 2.6.2.

Dmitri Aleksandrovich Grave was born August 25 (September 6), 1863
in Kirillov in the province of Vologda.[13] After the father's death the family
moved to St Petersburg, where he finished the grammar school with the gold
medal in 1881.

---

[11] Source: [Kuz40].
[12] The sources did not say clearly when he finished his lecturer's jobs at these places.
[13] Source: [Dob68].

From 1881 until 1885 Grave studied at the St Petersburg University; in 1885 he defended his candidate dissertation 'About minimal surfaces'. In 1888 there followed his master thesis 'About the integration of partial differential equations of first order' defended before Korkin and Sochocki. In his doctoral thesis 'About Fundamental Problems of the Mathematical Theory of the Construction of Geographic Maps' he gave the above-mentioned solution of a problem of Chebyshev among other results. The official opponents were here A. Markov und Ptashitski.

Since 1891 Grave gave lectures at the women's university, since 1893 additionally at the military-topographical institute. In 1899 he was appointed ordinary professor at Kharkiv University by initiative of Tikhomandritski. He worked there until 1902, when he moved to Kyiv as a successor of Pokrovski, who had died shortly before. Grave stayed in Kyiv until his death and taught at several institutes.

In 1920 Grave became the first mathematician who was elected ordinary member of the academy of sciences of the Ukraine, founded in 1919. In 1924 Grave was elected corresponding member of the academy of sciences of the USSR, 1929 honorary member.

In 1933 he became foundation rector of the mathematical institute of the academy of sciences of the Ukraine.

Dmitri Aleksandrovich Grave died December 9, 1939 in Kyiv.

A.9 Georgi Feodosievich Voronoy

In spite of his short life Georgi Feodosevich Voronoy counts as the most important representative of number theory of the St Petersburg Mathematical School after Chebyshev. The work 'Voronoï's Impact of Modern Science' published in 1998 by the Ukrainian Academy of Sciences is devoted to his work and life. An interesting fact is that this work is the first work of such a scope about a representative of the St Petersburg Mathematical School written not in Russian or another language of the Russian Empire and the Soviet Union, respectively. Even the Collected Works of Chebyshev, published in French in 1899, contained only uncommented texts.

Georgi Feodosevich Voronoy was born April 28, 1868 in the village of Zhuravki, district of Ryriatin, in the province of Poltava.[14]

In 1889 he finished his studies at the St Petersburg University, where in 1894 he defended his master thesis 'About integral algebraic numbers depending from the roots of an equation of third degree.'

In the same year Voronoy became professor at Warsaw University, where he defended his doctoral thesis 'About a generalization of a continuous fraction' in 1897.

Georgi Feodosevich Voronoy died after a severe disease in his home village November 20, 1908.

---

[14] Source: [Syt98].

B

Explanations

B.1 Russian Academic Degrees

It is not very simple to understand immediately what we are talking about, if we speak about a Russian academic degree. Therefore we put together a list of the most important names we had to use in the present work.

We emphasize that we do not want to judge the possibility of comparing scientific efforts by comparing degrees. Nowadays Russian degrees also have adopted the Anglo-American system of Bachelor and Master. So please have in mind that all such titles are flexible.

But nevertheless we think it becomes clear why Russian officials had difficulties to acknowledge foreign degrees.

adjunkt In the 19th name 'Adjunkt' (адъюнкт) was used also in the civil world in the sense of an assistant: at the academy of sciences and at the university. At the academy the ranking was: Adjunkt, extraordinary member and ordinary member.

candidate In prerevolutionary Russia the academic degree of the candidate (кандидат наук) was reached after the defense of the candidate thesis and passing the candidate examinations and was the first academic degree. It can be compared with the modern diploma or M.Sc. After the revolution of 1917 and the reorganisation of the universities it named the second academic degree, following the diploma, comparable with the Ph.D.

magister The degree of a magister (магистер наук) only existed before the revolution. It marked the degree between the candidate and the doctorate, so it could be compared with the Ph.D. in spite of its name.

magister-examination The magister examination had to be passed to get the right to write a magister thesis. The degree magister was then awarded after the defense of the thesis, but not after an additional examination.

pro venia legendi With the thesis pro venia legendi one got the right to teach at the university. It did not replace the doctoral thesis.

privat-dotsent A 'privat-dotsent' (приват-доцент) had the right and the duty to teach at the university without a salary. Lecturers who got a salary for their labor, at least had the status of a 'dotsent' (доцент). In Germany this system is still actual.

doctorate The degree of a 'doctor' (доктор наук) traditionally is the highest degree in Russia reached by an examination. It is awarded after the public defense of the doctoral thesis.

Bibliography

[Akh48]   AKHIEZER, NAUM ILICH Andrey Andreevich Markov. Biographical Notes [Андрей Андреевич Марков. Биографический очерк] *in:* [MarA48, p. 7-12].

[Akh55]   AKHIEZER, NAUM ILICH *The Academician S. N. Bernstein and His Contributions to Constructive Function Theory* [Академик С. Н. Бернштейн и его работы по конструктивной теории функций,] Изд. Харьковского государственного университета, Kharkov 1955.

[Aka45/1]   AKADEMIYA NAUK SSSR (HRSG.) *The Scientific Heritage of P. L. Chebyshev, vol. 1: Mathematics* [Научное наследие П. Л. Чебышева, вып. 1: математика], изд. АН СССР, Moscow-Leningrade 1945.

[Aka45/2]   AKADEMIJA NAUK SSSR (HRSG.) *The Scientific Heritage of P. L. Chebyshev, vol. 2: Mechanism Theory* [Научное наследие П. Л. Чебышева, вып. 2: теория механизмов], изд. АН СССР, Moscow-Leningrad 1945.

[Ale98]   ALEKSANDROV, PAVEL SERGEEVICH (ED.) *Hilbert's Problems,* Harri Deutsch, Thun / Frankfurt a. M. 1998.

[Amt10]   *Amtliches Verzeichnis des Personals und der Studierenden der Königlichen Georg-August-Universität zu Göttingen auf das halbe Jahr von Ostern bis Michaelis 1910,* Göttingen, Universitätsarchiv.

[AndL31]   *At the Leningrad Mathematical Front* [На Ленинградском математическом фронте,] государственное социально-экономическое издательство, Moscow / Leningrad 1931.

[Andr90-05]   ANDREEVSKI, IVAN EFIMOVICH ET AL. (EDS.) Encyclopedic Dictionary [Энциклопедический словарь,] Brockhaus/Efron, Leipzig/St Petersburg 1890–1905.

[ArLe45]   ARTOBOLEVSKI, IVAN IVANOVICH; LEVITSKI, NIKOLAY IVANOVICH The Mechanisms of Pafnuti Lvovich Chebyshev, [Механизмы П. Л. Чебышева] *in:* [Aka45/2, p. 7–109].

[ArLe55/1]   ARTOBOLEVSKI, IVAN IVANOVICH; LEVITSKI, NIKOLAY IVANOVICH The Models of the Mechanisms of Pafnuti Lvovich Chebyshev [Модели механизмов П. Л. Чебышева] *in:* [Chebgw3, vol. 2, p. 888-919].

[ArLe55/2]   ARTOBOLEVSKI, IVAN IVANOVICH; LEVITSKI, NIKOLAY IVANOVICH
Comments to the Articles about Mechanism Theory [Комментарии к
статьям по теории механизмов] *in:* [Chebgw3, vol. 2, S. 920-923].

[Bazh88]   BAZHANOV, VASILI ANATOLEVICH *Nikolay Aleksandrovich Vasilev*
[Николай Александрович Васильев], наука, Moscow 1988.

[Beri03/04]   *Bericht des Mathematischen Vereins an der Universität Göttingen
über sein LXX. Semester,* Göttingen 1904.

[Beri07]   *Bericht des Mathematischen Vereins an der Universität Göttingen
über sein LXXVII. Semester,* Göttingen 1910.

[BernHil]   BERNSTEIN, SERGEY NATANOVICH *6 Letters to David Hilbert 1905-
1910,* Niedersächsische Staats- und Universitätsbibliothek Göttingen,
Cod. Ms. D. Hilbert Nr. 23.

[Bern03]   BERNSTEIN, SERGEY NATANOVICH Sur la nature analytique des solutions de certaines équations aux dérivées partielles du second ordre,
Comptes Rendus **137** (1903), 778-781.

[Bern04/1]   BERNSTEIN, SERGEY NATANOVICH *Sur la nature analytique des solutions des équations aux dérivées partielles du second ordre,* Thèse fac.
sci., Paris 1904.

[Bern04/2]   BERNSTEIN, SERGEY NATANOVICH Sur la nature analytique des solutions des équations aux dérivées partielles du second ordre, *Math.
Annalen* **59** (1904), 20-76.

[Bern05]   BERNSTEIN, SERGEY NATANOVICH Sur l'interpolation, *Bulletin de la
Soc. Math. de France* **XXXIII** (1905), 33-36.

[Bern08]   BERNSTEIN, SERGEY NATANOVICH Investigation and Solution of Elliptic Partial Differential Equations of Second Order [Исследование
и интегрирование дифференциальных уравнений с частными
производными второго порядка эллиптического типа], *Сообщ.
Харьк. мат. общ-ва, серия 2* **11** (1908), 1-96.

[Bern11]   BERNSTEIN, SERGEY NATANOVICH Sur l'approximation des fonctions
continues par des polynomes, *Comptes Rendus* **152**, (1911), 502-504,
russian translation in: [Bern52, p. 8-10].

[Bern12/1]   BERNSTEIN, SERGEY NATANOVICH Démonstration du théorème de
Weierstrass fondée sur la calcul des probabilités, сообщ. Харьк.
матем. об-ва, серия 2, **13**, (1912), 1-2, *russian translation in:*
[Bern52, p. 105-106].

[Bern12/2]   BERNSTEIN, SERGEY NATANOVICH About Best Approximation of
Continuous Functions by Means of Polynomials of Given Degree
[О наилучшем приближении непрерывных функций посредством многочленов данной степени], *сообщ. Харьк. матем. об-ва,
серия 2,* **13**, (1912), 49-194, *in:* [Bern52, p. 11-104].

[Bern12/3]   BERNSTEIN, SERGEY NATANOVICH Sur la valeur asymptotique de la
meilleure approximation de $|x|$, *Comptes Rendus* **154**, (1912), 184-186.

[Bern12/4]   BERNSTEIN, SERGEY NATANOVICH Sur l'ordre de la meilleure approximation des fonctions continues par des polynomes de degré donné,
Mém. publ. par la classe des sciences Acad. de Belgique, sér. 2 **4**
(1912), 1-103.

[Bern13/1]   BERNSTEIN, SERGEY NATANOVICH Sur les recherches récentes relatives
à la meilleure approximation des fonctions continues par des polynomes, *in:* Proc. Int. Math. Congr., vol. 1, Cambridge 1912, 256-266.
russian translation in: [Bern52, p. 112-123].

[Bern13/2]   BERNSTEIN, SERGEY NATANOVICH Sur la meilleure approximation de $|x|$ par des polynomes de degrés donnés, *Acta mathematica* **37**, (1913), 1-57.

[Bern14]   BERNSTEIN, SERGEY NATANOVICH Sur la définition et les propriétés des fonctions analytiques d'une variable réelle, *Math. Annalen* **75** (1914), 449-468, *russian translation in:* [Bern52, p. 231-250].

[Bern16]   BERNSTEIN, SERGEY NATANOVICH Quelques remarques sur l'interpolation, сообщ. Харьк. матем. об-ва, серия 2, **15**, (1916), S. 49-62, *russian translation in:* [Bern52, p. 253-263].

[Bern26]   BERNSTEIN, SERGEY NATANOVICH Leçons sur les propriétés ex-tremales et la meilleure approximation des fonctions analytiques d'une variable réelle, *in: É. Borel (Hg.): Collection de monographies sur la théorie des fonctions,* tome 10, Gauthier-Villars, Paris 1926.

[Bern32]   BERNSTEIN, SERGEY NATANOVICH The Current State and the Problems of the Theory of Best Approximation of Real-Valued Functions by Polynomials [Сучасний стан та проблеми теорії найкращого наближення функції дійсної змінної через поліноми], *Зап. Харк. мат. тов. сер. 4* **5** (1932), 21-35.

[Bern37]   BERNSTEIN, SERGEY NATANOVICH *Extremal Properties of Polynomials and the Best Approximation of Continuous Functions of a Real Variable* [Экстремальные свойства полиномов и наилучшее приближение непрерывных функций одной вещественной переменной], ОНТИ, Moscow-Leningrad 1937.

[Bern45/1]   BERNSTEIN, SERGEY NATANOVICH The Academician Chebyshev (to the 50th anniversary of his death December 8, 1894) [Академик П. Л. Чебышева (к 50-летней годовщине его кончины 8 декабря 1894 г.)] *Природа* **3/1945**, 78-86 (equal word-for-word to [Bern47]).

[Bern45/2]   BERNSTEIN, SERGEY NATANOVICH Constructive Theory of Functions as the Development of Chebyshev's Ideas [Конструктивная теория функций, как развитие идей Чебышева,] *изв. АН СССР* **9** (1945), 145-158.

[Bern47]   BERNSTEIN, SERGEY NATANOVICH Chebyshev, His Influence on the Development of Mathematics [Чебышев, его влияние на развитие математики] *Ученые записки МГУ* **91** (1947), 35-45 (equal word-for-word to [Bern45/1]).

[Bern52]   BERNSTEIN, SERGEY NATANOVICH *Collected Works, Volume I: Constructive Theory of Functions* [Собрание сочинений, Том I: Конструктивная теория функций], изд. АН СССР, Moscow 1952.

[Bert64]   BERTRAND, JOSEPH *Traité de Calcul Différentiel et de Calcul Intégral, première partie: Calcul Différentiel,* Gauthier-Villars, Paris 1864.

[Bio96]   *Biographical Dictionary of the Professors and Lecturers of the Imperial St Petersburg University for the Past Third Quarter of a Century of its Existence* [Биографический словарь профессоров и преподавателей императорского Санкт Петербургского университета за третью четверть века его существования,], тип. Вольфа, St Petersburg, 1896.

[Bli01]   BLICHFELDT, H. F. Note on the Functions of the Form $f(x) \equiv \Phi(x) + a_1 x^{n-1} + a_2 x^{n-2} + \cdots + a_n$ which in a Given Interval Differ the least Possible Value from Zero, *Trans. Am. Math. Soc.* **2** (1901), 100-102.

204 Bibliography

[Bog91] BOGOLJUBOV, ALEKSEY NIKOLAEVICH Sergey Natanovich Bernstein [Сергей Натанович Бернштейн] *Вопросы истории естество-знания и техники* **3** (1991), 56-65.

[Bor05] BOREL, ÉMILE Leçons sur les Fonctions de Variables Réelles et les Développements en séries de Polynomes, Gauthier-Villars, Paris 1905.

[BoBu87] BORODIN, ALEKSEY IVANOVICH; BUGAY, ARKADI SILVESTROVICH *Prominent Mathematicians* [Выдающиеся математики], Радянська школа, Kyiv 1987.

[But92] BUTS, OLGA NIKOLAEVNA *The Development of Chebyshevian Approximation Theory* [Развитие Чебышевской теории приближения], Candidate Thesis, Kyiv 1992.

[BuJo89] BUTZER, PAUL L. AND JONGMANS, FRANÇOIS P. L. Chebyshev (1821-1894) and His Contacts with Western European Scientists, *Historia Mathematica* **16** (1989), 46-68.

[BuJo91] BUTZER, PAUL L. AND JONGMANS, FRANÇOIS Eugène Catalan and the rise of Russian science, *Bull. de l'académie royale de Belgique, Classe des Sciences* 6e série, t. II (1991), 59-90.

[BuJo99] BUTZER, PAUL L. AND JONGMANS, FRANÇOIS P. L. Chebyshev (1821-1894). A Guide to his Life and Work, *J. Appr. Theory* **96** (1999), 111-138.

[BuSt] BUTZER, PAUL L. AND STARK, EBERHARD L. The singular integral of Landau alias the Landau polynomials - Placement and impact of Landau's article: „über die Approximation einer stetigen Funktion durch eine ganze rationale Funktion", *in:* P. T. Bateman et al. (Eds.) *Edmund Landau. Collected Works,* vol. 3, without a date, S. 83-111.

[Chebgw1] CHEBYSHEV, PAFNUTI LVOVICH *Œuvres de P. L. Tchebychef,* Publiées par les soins de MM. A. Markoff et N. Sonin, Acad. Imp. des sciences, St. Petersburg 1899.

[Chebgw2] CHEBYSHEV, PAFNUTI LVOVICH *Complete Collected Works* [Полное собрание сочинений, изд. АН СССР,] 5 volumes, Moscow - Leningrad 1946–1951.

[Chebgw3] CHEBYSHEV, PAFNUTI LVOVICH *Selected Works* [Избранные труды, изд. АН СССР,] Moscow - Leningrade 1955.

[Chebgw4] CHEBYSHEV, PAFNUTI LVOVICH *Selected Mathematical Works* [Избранные математические труды], гостехиздат технико-теоретической литературы, Moscow - Leningrad 1946.

[Cheb38] CHEBYSHEV, PAFNUTI LVOVICH *Computation of the Roots of Equations* [Вычисление корней уравнений], Moscow 1838, *in:* [Chebgw2, vol. 5, p. 7-25].

[Cheb45] CHEBYSHEV, PAFNUTI LVOVICH *An Attempt to an Elementary Analysis of the Theory of Probabilities* [Опыт елементарного анализа теории вероятностей], Moscow 1845, *in:* [Chebgw2, vol. 5, p. 26-85].

[Cheb47] CHEBYSHEV, PAFNUTI LVOVICH About Integration with the help of Logarithms. Thesis pro venia legendi [Об интегрировании помощью логарифмов], *reprinted in the Изв. АН СССР*, **8**, 1930 *and in* [Chebgw2, vol. 5, p. 88-140].

[Cheb48] CHEBYSHEV, PAFNUTI LVOVICH Sur la fonction qui détermine la totalité des nombres premiers inférieure à une limite donnée *Mémoires des Savants Étrang. présentes à l'Academie Impériale des Sciences de St. Pétersbourg* **VI** (1848), also *Journal de Liouville* **17** (1852).

[Cheb49]   CHEBYSHEV, PAFNUTI LVOVICH *Theory of Congruences* [Теория сравнений] St Petersburg 1849, *German:* Theorie der Kongruenzen, Berlin 1888, *Italian:* Teoria delle congruenze, Rome 1895.

[ChebTrust]   CHEBYSHEV, PAFNUTI LVOVICH To the Trustee of the St Petersburg Educational District with a Report about the Official Trip to France [Письмо П. Л. Чебышева попечителю С-Петербургского учебного округа с отчетом о коммандировке во Францию], Letter from 3. 10. 1852, Арх. Лнгр. обл., Пб. университет; ф. 14, св. 81, д. 5а, л. 110-111, *in* [Chebgw2, vol. 5, p. 243-244]

[Cheb52/1]   CHEBYSHEV, PAFNUTI LVOVICH Report of the Extraordinary Professor Chebyshev about a Scientific Trip to England [Отчет экстраординарного профессора Чебышева о научной поездке в Англию], Letter without a date (about the end of 1852), Арх. Лнгр. обл., Пб. университет; ф. 14, св. 81, д. 5а, л. 112-113, *in:* [Chebgw2, vol. 5, p. 245].

[Cheb52/2]   CHEBYSHEV, PAFNUTI LVOVICH Report of the Extraordinary Professor of St Petersburg University Chebyshev about the Trip Abroad [Отчет экстраординарного профессора С-Петербургского университета Чебышева о путешествии за границу] Журнал министерства народного просвещения, ч. XXVIII, отд. IV, S. 2-14, *in:* [Chebgw2, vol. 5, p. 246-255].

[Cheb54]   CHEBYSHEV, PAFNUTI LVOVICH Théorie des mécanismes, connus sous le nom de parallélogrammes, *Mémoires présentes à l'Academie Impériale des Sciences de St. Pétersbourg par divers savants* **VII**, (1854), p. 539-568.

[Cheb55/1]   CHEBYSHEV, PAFNUTI LVOVICH Sur une formule d'Analyse, *Bull. de la classe phys.-math. de l'Acad. Imp. des Sciences de St.-Pétersbourg,* **XIII** (1855), also *Crelle* **53** and *in:* [Chebgw2, vol. II, S. 99-100].

[Cheb55/2]   CHEBYSHEV, PAFNUTI LVOVICH About Continuous Fractions [О непрерывных дробях,] *Уч. зап. Имп. Акад. наук,* III, 1855, S. 636-664, *in:* [Chebgw1, vol. I, p. 203-230] and [Chebgw2, vol. II, p. 103-126].

[Cheb56/1]   CHEBYSHEV, PAFNUTI LVOVICH Sur la construction des cartes géographiques *Bull. de la classe phys.-math. de l'Acad. Imp. des Sciences de St.-Pétersbourg,* **XIV** (1856), pp. 257-261, [Chebgw2, vol. 5, p. 146-149].

[Cheb56/2]   CHEBYSHEV, PAFNUTI LVOVICH *Drawing Geographical Maps* [Черчение географических карт], Сочинение, написанное для торжественного акта в Императорском С.-Петербургском университете 8 февраля 1856 г., St. Petersburg 1856, *in:* [Chebgw2, vol. 5, p. 150-157].

[Cheb57]   CHEBYSHEV, PAFNUTI LVOVICH Questions about Minima Connected with the approximative Representation of Functions (Short Note) [Вопросы о наименших величинах, связанные с приближенным представлением функций (Краткая заметка)], *Bull. de la classe phys.-math. de l'Acad. Imp. des Sciences de St.-Pétersbourg,* **XVI** (1857), p. 145-149 *in:* [Chebgw2, vol. 2, S. 146-150].

[Cheb58]   CHEBYSHEV, PAFNUTI LVOVICH Sur une nouvelle série, *Bull. de la classe phys.-math. de l'Acad. Imp. des Sciences de St.-Pétersbourg,* **XVII** (1858), pp. 257-261, [Chebgw2, vol. 2, p. 236-238].

[Cheb59]   CHEBYSHEV, PAFNUTI LVOVICH Sur les questions de minima qui se rattachent à la représentation approximative des fonctions, *Mém. de l'Acad. de St. Pétersbourg,* **série VI, t. VII,** (1859), p. 199-291.

[Cheb59/2]   CHEBYSHEV, PAFNUTI LVOVICH Sur le développement des fonctions à une seule variable, *Bull. de la classe phys.-math. de l'Acad. Imp. des Sciences de St.-Pétersbourg,* **I,** (1859) pp. 193-200, [Chebgw2, vol. 2, p. 335-341].

[Cheb61]   CHEBYSHEV, PAFNUTI LVOVICH Sur une modification du parallélogramme articulé de Watt, *Bulletin de l' Acad. de St. Pétersbourg,* **t. IV.,** *in:* [Chebgw2, vol. 4].

[Cheb69]   CHEBYSHEV, PAFNUTI LVOVICH A Rule for the Approximative Determination of Distances on the Surface of the Earth [Правило для приближенного определения расстояний на поверхности земли] *in:* Месяцеслове на 1869 год, изд. акад. наук, стр. 128, St. Petersburg 1869, [Chebgw2, vol. 2, p. 736].

[Cheb70]   CHEBYSHEV, PAFNUTI LVOVICH About Functions Similar to Legendre's [О функциях, подобных функциям Лежандра], *зап. Имп. Акад. наук, т.* **XVI,** 1870, 131-140, in: [Chebgw1, vol. 2, p. 61-68] und [Chebgw2, vol. 3, p. 5-12].

[Cheb71]   CHEBYSHEV, PAFNUTI LVOVICH About the Centrifugal Regulator [О центробежном уравнителе] отчет Имп. Московского технического училища за 1871 год, 1871, *in:* [Chebgw3, p. 689-709].

[Cheb73]   CHEBYSHEV, PAFNUTI LVOVICH About Functions Least Deviating from Zero[О функциях, наименее уклоняющихся от нуля], zu Band XXII der *Зап. Акад. Наук* **N0 1,** St. Petersburg 1873, [Chebgw2, vol. 2, p. 189-215].

[Cheb78]   CHEBYSHEV, PAFNUTI LVOVICH Sur la coupe des vêtements, Speech given before the Association française pour l'avancement des sciences, August 28, 1878, *Russian translation in:* [Chebgw2, vol. 5, p. 165-170].

[Cheb80]   CHEBYSHEV, PAFNUTI LVOVICH About Functions a Little Deviating from Zero under Certain Given Values for the Variables [О функциях, мало удаляющихся от нуля при некоторых величинах переменной], *Mém. de l'Acad. de St. Pétersbourg,* **LX,** St. Petersburg 1880, *in:* [Chebgw2, vol. 4].

[Cheb88]   CHEBYSHEV, PAFNUTI LVOVICH About the Simplest Joint Mechanism for the Creation of a Symmetric Motion round an Axis [О простейшей суставчатой системе, доставляющей движения, симметрические около оси], *Mém. de l'Acad. de St. Pétersbourg,* **LX,** St. Petersburg 1888, [Chebgw2, vol. 4].

[Cheb89]   CHEBYSHEV, PAFNUTI LVOVICH About Approximations of the Square Root of a Variable by Simple Fractions [О приближенных выражениях квадратного корня через простые дроби], Anlage zu Band LXI der *Зап. Акад. Наук* **№ 1,** St Petersburg 1889.

[Cheb93]   CHEBYSHEV, PAFNUTI LVOVICH About Polynomials Best Approximating the Simplest Rational Functions with their Variables Bounded Between Two Values [О полиномах, наилучше представляющих значения простейших дробных функций при величинах переменной, заключающихся между двумя данными пределами], Appendix to Volume LXXII of the *Зап. Акад. Наук* **№ 7,** St Petersburg 1893.

[Che66]   CHENEY, E. W. *Introduction to Approximation Theory*, McGraw-Hill, New York, 1966.

[ChLo66]   CHENEY, E. W. AND LOEB, H. L. On the continuity of rational approximation operators, *Arch. Rational Mech. Anal.* **21** (1966), 391-401.

[Del00]   DELONE, NIKOLAJ BORISOVICH Die Tschebyschefschen Arbeiten in der Theorie der Gelenkmechanismen, *in:* [VaDe00]

[Delo99]   *The Affair of the Academician Nikolay Nikolaevich Luzin* [Дело академика Николая Николаевича Лузина,] Русский христианский гуманитарный институт, St. Petersburg 1999.

[Del45]   DELONE, BORIS NIKOLAEVICH *The Academician Pafnuti Lvovich Chebyshev and the Russian Mathematical School* [Академик Пафнутий Львович Чебышев и русская школа математиков,] Speech on the celebrating session of the anniversary year of the academy of sciences of the USSR, изд. АН СССР, Moscow - Leningrad, 1945.

[Dob68]   DOBROVOLSKI, VYACHESLAV ALEKSEEVICH *Dmitri Aleksandrovich Grave. 1863-1939* [Дмитрий Александрович Граве], наука, Moscow 1968.

[DuB76]   DU BOIS-REYMOND, P. Untersuchungen über die Convergenz und Divergenz der Fourierschen Darstellungsformeln, *Abhandlungen der Mathematisch-Physicalischen Classe der K. Bayerischen Akademie der Wissenschaften* **12** (1876), 1-103.

[EgoHil]   EGOROV, DMITRI F. *Letter from February 19, 1906 to David Hilbert*, Niedersächsische Staats- und Universitätsbibliothek Göttingen, Cod. Ms. D. Hilbert 90, Nr. 3.

[Erm87]   ERMOLAEVA, NATALIYA SERGEEVNA About a Non-Published Lecture about Probability Theory of P. L. Chebyshev [Об одном неопубликованном курсе теории вероятностей П. Л. Чебышева], *вопроси истории естествознания и техники* **4** (1987), p. 106-112.

[Erm94/1] ERMOLAEVA, NATALIYA SERGEEVNA Petersburg Mathematicians and the Theory of Analytical Functions [Петербургские математики и теория аналитических функцций], *историко-математические исследования* **35** (1994) p. 23-55.

[Erm94/2] ERMOLAEVA, NATALIYA SERGEEVNA On the History of the St. Petersburg and Petrograd Mathematical Societies, *Amer. Math. Soc. Transl.* **159,** (1994), 213-221.

[Erm96]   ERMOLAEVA, NATALIYA SERGEEVNA Julian Vasilevich Sochocki [Юлиан Васильевич Сохоцкий], *труды Санкт-Петербургского матем. общ.* **4** (1996) p. 359-364.

[Erm97]   ERMOLAEVA, NATALIYA SERGEEVNA The Finite and Infinite in the work of P. L. Chebyshev [Конечное и бесконечное в трудах П. Л. Чебышева], *in:* BARABASHEVA, A. G. (RED.): *Infiniteness in Mathematics: Philosophical and Historical Aspects* [Бесконечность в математике: философские и исторические аспекты], Янус-К, Moscow 1997.

[Erm98/1] ERMOLAEVA, NATALIYA SERGEEVNA (JERMOŁAJEWA, NATALIA S.) Julian Karol Sochocki - Uzupełnienie biografii naukowej, *Uniwersytet Szczeciński - materiały, konferencje* **30,** (1998), 47-65.

[Erm98/2]   ERMOLAEVA, NATALIYA SERGEEVNA About the So-Called 'Leningrad Mathematical Front' [О так называемом «Ленинградском математическом фронте»], *труды Санкт-Петербургского матем. общ.* **5** (1998) S. 380-395.

[Eul47]   EULER, LEONHARD *Introductio in analysin infinitorum,* Lausanne 1747 *(also in: Leonhardi Euleri opera omnia,* ser. I, vol. VIII, Teubner, Leipzig/Berlin 1922).

[Eul55]   EULER, LEONHARD *Institutiones calculi differentialis,* St. Petersburg 1755 *(also in: Leonhardi Euleri opera omnia,* ser. I, vol. X, Teubner, Leipzig/Berlin 1922).

[Eul75]   EULER, LEONHARD *Commercium Epistolicum,* ediderunt A. P. Jushkevich, V. I. Smirnov und W. Habicht, Opera Omnia Series IVa, Vol. 1, Birkhäuser, Basel 1975.

[Eul77]   EULER, LEONHARD De proiectione geographica De Lisliana in mappa generali imperii Russici usitata, *Acta academiae scientiarum Petropolitanae* **I, 1178** (1777), p. 143-153 (also *in:* Leonhardi Euleri opera omnia, ser. I, vol. XXVIII, Lausanne 1955).

[Fab14]   FABER, G. über die interpolatorische Darstellung stetiger Funktionen, *Jahresber. Deut. Math. Verein.* **23** (1914), 190-210.

[Fej00]   FEJÉR, LIPÓT Sur les fonctions bornées et intégrables, *Comptes Rendus de l'Académie des Sciences* **131** (1900), 984-987, *also in:* [Fej70, vol. 1, p. 37-41].

[Fej16/1]   FEJÉR, LIPÓT Interpolatióról (Első közlemény), *Mat. és Term. Értesítő* **34** (1916), 209-229, *also in:* [Fej70, vol. 2, p. 9-25].

[Fej16/2]   FEJÉR, LIPÓT über Interpolation, *Göttinger Nachrichten* **1** (1916), 66-91, *also in:* [Fej70, vol. 2, p. 25-48].

[Fej70]   FEJÉR, LIPÓT *Gesammelte Arbeiten,* 2 Bände (Pál Turán, ed.), Birkhäuser, Basel/Stuttgart 1970.

[Fel60]   FEL', SERGEJ EFIMOVICH *Russia's Cartography in the* 18th *Century* [Картография России XVIII века], изд. географической литературы, Moscow 1960.

[Fis78]   FISHER, STEPHEN D. Quantitative Approximation Theory, *Amer. math. monthly* **85** (1978), 318-332.

[Fou22]   FOURIER, JEAN BAPTISTE JOSEPH *Théorie analytique de la chaleur,* Didot, Paris 1822 (auch *in: Œuvres de Fourier, publiées par les soins de M. Gaston Darboux,* t. 1, Gauthier-Villars, Paris 1888)

[Fou31]   FOURIER, JEAN BAPTISTE JOSEPH *Analyse des équations determinées,* Didot, Paris 1831.

[Fre07]   FRÉCHET, MAURICE Sur l'approximation des fonctions par des suites trigonométrques limitées, *Comptes Rendus de l'Académie des Sciences* **144** (1907), 124-125.

[Fre08]   FRÉCHET, MAURICE Sur l'approximation des fonctions continues périodiques par les sommes trigonométrques limitées, *Annales de l'École normale Supérieure* **25** (1908), 43-56.

[Ger54]   GERONIMUS, JA. L. *Pafnuti Ljwowitsch Tschebyschew (1821-1894). Lösung kinematischer Probleme durch Näherungsmethoden* VEB Verlag Technik, Berlin 1954.

[Gon45]   GONCHAROV, VASILI LEONIDOVICH Theory of Best Approximation [Теория наилучшего приближения функций] *in:* [Aka45/1, p. 122-172].

[Gon47/1]   GONCHAROV, VASILI LEONIDOVICH Theory of Mechanisms Known under the Name of Parallelograms (Comment) [Теория механизмов, известных под названием параллелограммов (Комментарий)], *in:* [Chebgw2, vol. 2, p. 474-485].

[Gon47/2]   GONCHAROV, VASILI LEONIDOVICH Questions about Minima Connected with the approximative Representation of Functions (Comment) [Вопросы о наименьших величинах, связанные с приближенным представлением функций (Комментарий)], *in:* [Chebgw2, vol. 2, p. 496-507].

[Gra96]   GRAVE, DMITRI ALEKSANDROVICH *About Fundamental Problems of the Mathematical Theory of the Construction of Geographic Maps* [Об основных задачах математической теории построения географических карт], doctoral thesis, St Petersburg 1896.

[Gra11]   GRAVE, DMITRI ALEKSANDROVICH Démonstration d'un théorème de Tchébycheff généralisé, *Crelle,* **140** (1911).

[GrHo75]   GRÖBNER, WOLFGANG; HOFREITER, NIKOLAUS *Integraltafeln* (Erster Teil, unbestimmte Integrale), Wien/New York 1975.

[Gro87]   GRODZENSKI, SERGEY YAKOVLEVICH *Andrey Andreevich Markov. 1856-1922* [Андрей Андреевич Марков. 1856-1922], наука, Moscow 1987.

[Gus57]   GUSAK, ALEKSEY ADAMOVICH Application of the 'Method of Corrections' of the Solution of the Problem about the Best Construction of Watt's Parallelogram [Прымяненне ,,метаду направак" Чэбышэва да рашэння задачы аб найлепшай будове паралелаграма Уата], *весці Акадэміі навук Беларускай ССР, серыя фізіка- технічных навук,* **4** (1957), 73-83.

[Gus61]   GUSAK, ALEKSEY ADAMOVICH Prehistory and Initial Development of Approximation Theory [Предыстория и начало развития теории приближения функций] *историко-математические исследования* **14** (1961), 289-348.

[Gus72]   GUSAK, ALEKSEJ ADAMOVICH *Approximation Theory* [Теория приближения функций], изд. БГУ, Minsk 1972.

[Haa17]   HAAR, ALFRED Die Minkowskische Geometrie und die Annäherung an stetige Funktionen, *Mathematische Annalen* **78** (1917), 294-311.

[Har48]   HART, WILLIAM L. Dunham Jackson. 1888-1946, *Bull. Amer. Math. Soc.* **54** (1948), 847-860.

[Her64]   HERMITE, CHARLES Sur un nouveau développement en série de fonctions *Comptes Rendues* **58** (1864).

[Her78]   HERMITE, CHARLES Sur la formule d'interpolation de Lagrange, *Crelle* **84** (1878), 70-79.

[Jack11]   JACKSON, DUNHAM *Über die Genauigkeit der Annäherung stetiger Funktionen durch ganze rationale Funktionen gegebenen Grades und trigonometrische Summen gegebener Ordnung,* Dissertation, Göttingen 1911.

[Jush68]   JUSHKEVICH, ADOLF PAVLOVICH *History of Mathematics in Russia until 1917* [История математики в России до 1917 года], Наука, Moscow 1968.

[KayoJ]   KAYSERLICHE ACADEMIE DER WISSENSCHAFTEN *Rußischer Atlas welcher in einigen zwanzig Special-Charten das gesamte Rußische Reich mit den angräntzenden Ländern, zu desto gründlicher Verfertigung einer General-Charte dieses grossen Kayserthums nach den Regeln der Erdbeschreibung und den neusten Observationen vorstellig macht, entworffen bey der Kayserl. Academie der Wissenschaften*, St Petersburg, without a date (about 1741-45).

[KiHi] KIRCHBERGER, PAUL *Letter to David Hilbert 1922*, Niedersächsische Staats- und Universitätsbibliothek Göttingen, Cod. Ms. D. Hilbert 452 b): 34.

[Kir02] KIRCHBERGER, PAUL *Ueber Tchebychefsche Annäherungsmethoden*, Dissertation, Göttingen 1902.

[Kir03] KIRCHBERGER, PAUL über Tchebyschefsche Annäherungsmethoden, *Math. Annalen* **57** (1903), 509-540.

[Kle28] KLEIN, FELIX *Elementarmathematik vom höheren Standpunkt aus. Band III: Präzisions- und Approximationsmathematik,* 3. Auflage, Springer, Berlin 1928.

[Kor96] KORKIN, ALEKSANDR NIKOLAEVICH Sur les équations differentielles ordinaires du premier ordre, *Math. Annalen* **48** (1896), 317-364.

[KoZo73] KORKIN, ALEKSANDR NIKOLAEVICH; ZOLOTAREV, EGOR IVANOVICH Sur un certain minimum, *Nouvelles Annales de mathématiques, 2-e série* **12** (1873), 337-355, *in:* [Zol32, vol. 1, p. 138-153].

[Kry50] KRYLOV, ALEKSEY NIKOLAEVICH A Short Biographical Outline about A. N. Korkin [Краткий биографический очерк А. Н. Коркина] *in: Memories and Remarks* [Воспоминания и очерки,] изд. АН СССР (1950), 413-415.

[Kuz40] KUZMIN, RODION OSIEVICH Ivan Ivanovich Ivanov. 1862-1939 [Иван Иванович Иванов. 1862-1939], *izv. AN SSSR, ser. mat.* **4, No. 4-5** (1940), 357-359.

[Lag79] LAGUERRE, E. Sur l'integrale $\int_x^\infty \frac{e^{-x}}{x}\, dx$ *Bull. de la Soc. Math. Fr.* **7** (1879).

[Lap43] LAPLACE, PIERRE SIMON MARQUIS DE Traité de la mécanique céleste, *in: Œuvres de Laplace, vol. 2,* imprimerie royale, Paris 1843.

[Lan08] LANDAU, EDMUND Über die Approximation einer stetigen Funktion durch eine ganze rationale Funktion, *Rendiconti del Circolo Matemático di Palermo* **XXV** (1908), 337-345.

[Leb98] LEBESGUE, HENRI Sur l'approximation des fonctions, *Bulletin de la Soc. Math. de France* **XXII** (1898).

[Leb08] LEBESGUE, HENRI Sur la représentation approchée des fonctions, *Rendiconti del Circolo Matemático di Palermo* **XXVI** (1908), 325-328.

[Leb10/1] LEBESGUE, HENRI Sur les intégrales singulières, *Annales de la Faculté de Toulouse* Ser. **3,** vol. **I** (1910), 25-117, *in:* Henri Lebesgue, Œuvres scientifiques, Vol. 3, p. 259-351.

[Leb10/2] LEBESGUE, HENRI Sur la représentation trigonométrique approchée des fonctions satisfaisant à une condition de Lipschitz, *Bulletin de la Soc. Math. de France* **XXXVIII** (1910), 184-210

[LeTr] LENZ, EMIL, BUNJAKOVSKIJ, V. JA. AND SOMOV, O. I.) *To His Excellency, the Trustee of the St Petersburger Educational District - Application* [Его превосходительству, господину попечителю С-Петербургского учебного округа - представление], Brief vom 18. 12. 1850, Арх. Лнгр. обл., канцелярия попечителя СП. уч.

округа; ф. 139, св. 107, д. 5459, л. 1-2, *in:* [Chebgw2, vol. 5, p. 242-243].

[Lya95]   LYAPUNOV, ALEKSANDR MICHAYLOVICH Pafnuti Lvovich Chebyshev [Пафнутий Львович Чебышев], *сообщения Харьковского математического общества, 2-я серия,* т. **IV, № 5 и 6,** (1895), *in:* [Chebgw4, p. 9-21].

[Lor97]   LORENTZ, RUDOLPH A. (ED.) *George G. Lorentz. Mathematics from Leningrad to Austin. Selected Works in Real, Functional and Numerical Analysis,* 2 volumes, Birkhäuser, Boston/Basel/Berlin 1997.

[Mai56]   MAIRHUBER, J. C. On Haar's theorem concerning Chebychev approximation problems having unique solutions, *Proc. Amer. Math. Soc.* **7** (1956), 609-615.

[MarAKl]   MARKOV, ANDREY ANDREEVICH, *10 Letters to Felix Klein 1880-1901 (1 without a date),* Niedersächsische Staats- und Universitätsbibliothek Göttingen, Cod. Ms. F. Klein 10, 917-926.

[MarAoJ]   MARKOV, ANDREY ANDREEVICH *Differential Calculus* [Дифференциальное исчисление,] St Petersburg, without a date.

[MarA84]   MARKOV, ANDREY ANDREEVICH Determination of a Function with Respect to the Condition to Deviate as Little as Possible from Zero [Определение некоторой функции по условию наименее уклоняться от нуля], *сообщения и протоколы заседании математического общества при императорском Харьковском университете,* Kharkov 1884, p. 83-92.

[MarA88]   MARKOV, ANDREY ANDREEVICH *Introduction to Analysis* [Введение в анализ,] Lectures of Professor A. Markov in the Academic Year 1886/87, St Petersburg 1888.

[MarA90]   MARKOV, ANDREY ANDREEVICH About a Question of D. I. Mendeleev [Об одном вопросе Д. И. Менделеева], *зап. акад. наук,* **62** (1890).

[MarA98]   MARKOV, ANDREY ANDREEVICH About Extremal Values of Integrals Connected with the Interpolation Problem [О предельных величинах интегралов в связи с интерполированием], *зап. акад. наук, физ.-мат. отд.* т. **4, № 5** (1898), *in:* [MarA48, p. 146-230].

[MarA03]   MARKOV, ANDREY ANDREEVICH About an Algebraic Theorem of Chebyshev [Об одной алгебраической теореме Чебышева], *изв. акад. наук,* 1903.

[MarA06]   MARKOV, ANDREY ANDREEVICH *About Functions Least Deviating from Zero* [О функциах, наименее уклоняющихся от нуля], arranged after lectures of the member of the academy A. A. Markov, St Petersburg 1906.

[MarA48]   MARKOV, ANDREY ANDREEVICH, *Selected Works about the Theory of Continuous Fractions and the theory of Functions Least Deviating from Zero* [Избранные труды по теории непрерывных дробей и теории функций, наименее уклоняющихся от нуля], госиздат технико-теоретической литературы, Moscow-Leningrad 1948.

[MarV92]   MARKOV, VLADIMIR ANDREEVICH *About Functions Deviating the Least Possible from Zero on a Given Interval* [О функциах, наименее уклоняющихся от нуля в данном промежутке], типография имп. акад. наук, St Petersburg 1892.

[MarV97]   MARKOV, VLADIMIR ANDREEVICH *About Positive Quadratic Forms of Three Variables* [О положительных троичных квадратичных формах], Master thesis, St Petersburg 1897.

[MarV16]   MARKOV, VLADIMIR ANDREEVICH Über Funktionen, die auf einem gegebenen Intervall möglichst wenig von Null abweichen, *Math. Annalen* **LXXVII** (1916) p. 213-258.

[MathGö]   MATHEMATISCHE GESELLSCHAFT ZU GÖTTINGEN, *Protokollbuch Nr. 1*, Niedersächsische Staats- und Universitätsbibliothek Göttingen, Cod. Ms. Math. Archiv 49:1.

[MeSc64]   G. MEINARDUS AND D. SCHWEDT, Nichtlineare Approximationen, *Arch. f. Rat. Mech. and Anal.* **17** (1964), 297-326.

[Men01]   MENDELEEV, DMITRI IVANOVICH, *Notes about People's Education in Russia* [Записки о народном просвещении в Росии,] St Petersburg 1901.

[Mér96]   MÉRAY, C. Noveaux exemples d'interpolations illusoires, *Bull. Sci. Math.* **20** (1896), 266-270.

[MlKl]   MLODZEEVSKI, BOLESLAV KORNELIEVICH, *2 Letters to Felix Klein 1891/1905*, Niedersächsische Staats- und Universitätsbibliothek Göttingen, Cod. Ms. F. Klein 10.

[Nat49]   NATANSON, ISIDOR PAVLOVICH *Construcitve Function Theory* [Конструктивная теория функций], госиздат технико-теоритической литературы, Moscow-Leningrad 1949, *German:* Konstruktive Funktionentheorie, Akademie-Verlag, Berlin 1955.

[Ozhi66]   OZHIGOVA, ELENA PETROVNA *Egor Ivanovich Zolotarev* [Егор Иванович Золотарев], наука, Moscow-Leningrad 1966.

[Ozhi68]   OZHIGOVA, ELENA PETROVNA *Aleksandr Nikolaevich Korkin* [Александр Николаевич Коркин], наука, Leningrad 1968.

[Pau97]   PAUL, SIEGFRIED *Die Moscower mathematische Schule um N. N. Lusin. Entstehungs- und Entwicklungsgeschichte, Arbeitsprinzipien, Zerfall: Unter besonderer Berücksichtigung der Kampagne gegen Lusin im Sommer 1936*, Kleine, Bielefeld 1997.

[Pic04]   PICARD, HENRI Rapport sur la thèse de S. Bernstein, Sur la nature analytique des solutions des équations aux dérivées partielles du second ordre, soutenu le 10 Juin 1904, *in:* Hélène Gispert, *La France mathématique. La Société mathématique de France (1872-1914)*, p. 389-390.

[Pin99]   PINKUS, ALLAN *Weierstrass and Approximation Theory*, Draft Dec. 1999.

[Pon35]   PONCELET, JEAN VICTOR Sur la valeur approchée lineaire et rationelle des radicaux de la forme $\sqrt{a^2 + b^2}$, $\sqrt{a^2 - b^2}$ etc., *Crelle*, **13** (1835), 277-292.

[PoKl]   POSSE, KONSTANTIN ALEKSANDROVICH *1 Letter to Felix Klein*, Niedersächsische Staats- und Universitätsbibliothek Göttingen, Cod. Ms. F. Klein 11, 371.

[Pos73]   POSSE, KONSTANTIN ALEKSANDROVICH *About Functions Similar to Legendre's* [О функциях, подобных функциям Лежандра,] Master Thesis, St Petersburg 1873.

[Pos86]   POSSE, KONSTANTIN ALEKSANDROVICH *Sur quelques applications des fractions continues algébriques*, St Petersburg 1886.

[Pos91]   POSSE, KONSTANTIN ALEKSANDROVICH *Lectures about Differential Calculus* [Лекции дифференциального исчисления,] Lectures, given by professor K. Posse. St Petersburg University, 1890/1.

[Pos03]   POSSE, KONSTANTIN ALEKSANDROVICH *Lectures about Calculus* [Курс дифференциального и интегрального исчислений,] put together by professor K. Posse, St Petersburg, 1903.

[Pos04]   POSSE, KONSTANTIN ALEKSANDROVICH Pafnuti Lvovich Chebyshev. A Short Biographical Outline [Пафнутий Львович Чебышев. Краткий биографический очерк], *in:* VENGEROV, S. A.: *Critical-Biographical Dictionary [Критико-биографический словарь]* vol. 6, St Petersburg 1904, p. 14-15, also in [Chebgw2, vol. 1, p. 5-9].

[Pos09]   POSSE, KONSTANTIN ALEKSANDROVICH Aleksandr Nikolaevich Korkin [Александр Николаевич Коркин,] *Мат. сборник* **27** (1909), 1-27.

[Pos15]   POSSE, KONSTANTIN ALEKSANDROVICH Nikolay Yakovlevich Sonin [Николай Яковлевич Сонин,] *Сообщ. Харьковского мат. общ.* **XIV, 6** (1915), 275-293.

[Pru50]   PRUDNIKOV, VASILIJ EFIMOVICH *P. L. Chebyshev - Scientist and Teacher* [П. Л Чебышев - ученый и педагог], гос. учебно-педагог. изд. мин. просвещ. РСФСР, Moscow 1950.

[Pru76]   PRUDNIKOV, VASILIJ EFIMOVICH *Pafnuti Lvovich Chebyshev. 1821-1894* [Пафнутий Львович Чебышев. 1821-1894], Наука, Leningrad 1976.

[Psh06]   PSHEBORSKI, ANTONI-BONIFATSI PAVLOVICH Report of the Privat-Dotsent A. P. Psheborski about the Official Trip Abroad [Отчет о заграничной командировке приват-доцента А. П. Пшеборского], *зап. имп. Харьк. уни-тета*, **1** (1906), 26-34.

[Psh13/1]   PSHEBORSKI, ANTONI-BONIFATSI PAVLOVICH Report about the thesis of S. N. Bernstein 'About the Best Approximation of Continuous Functions by Means of Polynomials of Given Degree' [Отзыв о диссертации С. Н. Бернштейна ,,О наилучшем приближении непрерывных функций посредством многочленов данной степени”], *зап. имп. Харьк. уни-тета*, **4** (1913), 27-34.

[Psh13/2]   PSHEBORSKI, ANTONI-BONIFATSI PAVLOVICH About Some Polynomials Least Deviating from Zero on a Given Interval [О некоторых полиномах, наименее уклоняющихся от нуля в данном промежутке], *сообщ. Харьк. мат. общ-ва, вторая серия*, **XIV, № 1-2** (1913), 65-80.

[Psh13/3]   PSHEBORSKI, ANTONI-BONIFATSI PAVLOVICH (PCHÉBORSKI, A.) Sur quelques polynomes qui s'écartent le moins possible de zéro dans un intervalle donné, *Comptes Rendus* **156** (1913), 531-533.

[Psh14]   PSHEBORSKI, ANTONI-BONIFATSI PAVLOVICH (PCHÉBORSKI, A.) Sur une généralisation d'un problème de Tschébischeff et de Zolotareff, *Comptes Rendus* **158** (1914), 619-621.

[Ric85]   RICHENHAGEN, GOTTFRIED *Carl Runge (1856-1927): Von der reinen Mathematik zur Numerik,* Vandenhoeck & Ruprecht, Göttingen 1985.

[Run85/1]   RUNGE, CARL Zur Theorie der eindeutigen analytischen Funktionen, *Acta mathematica* **VI** (1885), 229-244.

[Run85/2]   RUNGE, CARL Über die Darstellung willkürlicher Funktionen, *Acta mathematica* **VII** (1885), 387-392.

[Run01]   RUNGE, CARL Über empirische Funktionen und die Interpolation zwischen äquidistanten Ordinaten, *Zeitschrift für Mathematik und Physik* **46** (1901), 224-243.

[Run04]   RUNGE, CARL *Theorie und Praxis der Reihen* Göschen, Leipzig 1904.

[Run15]   RUNGE, CARL *Graphische Methoden* Teubner, Leipzig 1915 (2. Auflage Teubner, Leipzig/Berlin 1919).

[Sal67]   SALISTSCHEW, KONSTANTIN ALEKSEEVICH *Einführung in die Kartographie*, 2 Bände, VEB Geographisch-kartographische Anstalt Hermann Haack, Gotha/Leipzig 1967.

[Ser97]   SERGEEV, ALEKSANDR ANATOLEVICH *Konstantin Aleksandrovich Posse* [Константин Александрович Поссе], наука, Moscow 1997.

[She94]   SHEYNIN, O. Chebyshev's Lectures on the Theory of Probability *Arch. History Exact Sci.* **46** (1994), 321-340.

[Sib09]   SIBIRANI, FILIPO, Sulla rappresentazione approssimata delle funzioni, *Annali di Matematica Pura ed Applicata, ser. III* **16** (1909), 203-221.

[Smi53]   SMIRNOV, V. I. *An Outline of A. M. Ljapunov's Life* [Очерк жизни А. М. Ляпунова,] *in:* Akademie der Wissenschaften der UdSSR (Hrsg.) *Aleksandr Mikhaylovich Lyapunov. Bibliography* [Александр Михайлович Ляпунов. Библиограпхия, Изд. АН СССР, Moscow-Leningrad 1953.

[Son54]   SONIN, NIKOLAY YAKOVLEVICH *Investigations about cylindric functions and special polynomials* [Исследования о цилиндрических функциях и специальных полиномах,] госиздат технико-теоретической литературы, Moscow 1954.

[Stef94]   STEFFENS, KARL-GEORG *Über Alternationskriterien in der Geschichte der Besten Chebyshev-Approximation*, Diploma thesis, Frankfurt a. M. 1994.

[Stef99]   STEFFENS, KARL-GEORG Letters from S. N. Bernshteyn to David Hilbert, *East Journal on Approximations* **Vol 5, Nr. 4,** 1999, 501-515.

[Sue87]   SUETIN, PAVEL KONDRATEVICH *The Work of the Academician Andrey Andreevich Markov on the Subject of Analysis* [Работы академика А. А. Маркова по математическому анализу] *in:* [Gro87, p. 164-222].

[Sve01]   SVESHNIKOV, P. I. About Polynomials of Second, Third and Fourth Degree Least Deviating from Zero [О многочленах второй, третьей и четвертой степени, наименее уклоняющихся от нуля], *Журн. мин. нар. просвещ.* **1** (1901), 29-38.

[Syt98]   SYTA, HALYNA M. Short Biography of G. Voronoï *in: Voronoï's Impact on Modern Science, vol. 1,* NAS Ukraine, Kiev, 1998, 11-24.

[TiKl]   TIKHOMANDRITSKI, MATVEY ALEKSANDROVICH *7 Letters to Felix Klein 1883-1900,* Niedersächsische Staats- und Universitätsbibliothek Göttingen, Cod. Ms. F. Klein 12, 21-27.

[Tok87]   TOKARENKO, A. M. The Contribution of N. B. Delone to the Development of Mechanism Theory [Вклад Н. Б. Делоне в развитие теории механизмов] *in:* A. N. Bogoljubov (Responsible Red.): *Mathematical Recognition of Nature in its Development* [Математическое естествознание в его развитии,] Наукова думка, Kiev 1987, p. 153-156.

[Ton08]   TONELLI, LEONIDA I polinomi d'approssimazione di Tchebychev, *Annali di Matematica pura et applicata, serie III* **15** (1908), 47-119.

[Tra92]   TRAVCHETOV, I. I. About the Coefficients of the Parabola $px^2 + qx + r$ [О коэффициентах трехчлена] $px^2 + qx + r$, *Вестн. опытной физики и элем. матем.* **123** (1892), 59-63 and **127** (1892), 137-142.

[TrCo]   THE TRUSTEE *To the Council of St Petersburg University* [Совету С-Петербургского университета], Letter from29. 7. 1852, Арх. Лнгр. обл., Пб. университет; ф. 14, св. 81, д. 5а, л. 106, *in:* [Chebgw2, vol. 5, p. 243].

[Val08/1] VALLÉE-POUSSIN, CHARLES-JEAN DE LA Sur l'approximation des fonctions d'une variable réelle et de leurs dérivées par des polynômes et des suites limités de Fourier, *Bull. de l'académie royale de Belgique, Classe des Sciences* (1908), 193-254.

[Val08/2] VALLÉE-POUSSIN, CHARLES-JEAN DE LA Note sur l'approximation par un polynôme d'une fonction dont la dérivée est à variation bornée. *Bull. de l'académie royale de Belgique, Classe des Sciences* (1908), 403-410.

[Val10]   VALLÉE-POUSSIN, CHARLES-JEAN DE LA Sur les polynomes d'approximation et la représentation approchée d'un angle, *Bull. de l'académie royale de Belgique, Classe des Sciences* (1910), 808-844.

[Val19]   VALLÉE-POUSSIN, CHARLES-JEAN DE LA *Leçons sur l'approximation des fonctions d'une variable réelle,* Gauthier-Villars, Paris 1919.

[Vas00]   VASIL'EV, ALEKSANDR VASIL'EVICH P. L. Tschebyschef und seine wissenschaftlichen Leistungen *in:* [VaDe00].

[VaDe00]   VASILEV, ALEKSANDR VASILEVICH, DELONE, NIKOLAY BORISOVICH *P. L. Tschebyschef,* Teubner, Leipzig 1900.

[VaVv49-57] VAVILOV, S. I.; VVEDENSKIJ, B. A. (HRSG.) *Great Soviet Encyclopedia* [Большая советская энциклопедия,] 50 Bände, гос. научн. изд. «Большая советская энциклопедия», Moscow 1949-1957.

[Ver04]   *Verhandlungen des dritten internationalen Mathematiker-Kongresses in Heidelberg vom 8.-13. August 1904,* Teubner, Leipzig 1905.

[Vin77]   VINOGRADOV, IVAN MATVEEVICH (verantw. Red.) *Mathematical Encyclopedia* [Математическая Энциклопедия], Советская Энциклопедиа, Moscow 1977-1985.

[Vol87]   VOLKERT, KLAUS *Geschichte der Analysis,* Bibliographisches Institut, Mannheim / Wien / Zürich 1987.

[VorSPb]   *University Calendar of the Mathematico-Physical Faculty of the Imperial St Petersburg University,* [Обозрение преподавания наук в императорском С-Петербургском университете], Gorki-library, St Petersburg.

[Wal31]   WALSH, J. L. The existence of rational functions of best approximation, *Trans. Am. Math. Soc.* **33,** (1931), S. 668-689.

[Wei85]   WEIERSTRASS, KARL *Über die analytische Darstellbarkeit sogenannter willkürlicher Funktionen einer reellen Veränderlichen,* Sitzungsberichte der Akademie zu Berlin, Berlin 1885, S. 633-639.

[You07]   YOUNG, JOHN WESLEY General Theory of Approximation by Functions Involving a Given Number of Arbitrary Parameters, *Trans. Am. Math. Soc.* **23,** (1907), S. 331-344.

[Zol68]   ZOLOTAREV, EGOR IVANOVICH *About one Question on Minima* [Об одном вопросе о наименьших величинах], Diss. pro venia legendi, St. Petersburg 1868, *in:* [Zol32, вып. 2, p. 130-166].

[Zol70]   ZOLOTAREV, EGOR IVANOVICH *Differential Calculus* [Дифференциальное исчисление,] Lithographic notes of the lectures for students of the course of studies 'natural sciences' of the physico-mathematical faculty, St Petersburg 1870.

[Zol77/1]   ZOLOTAREV, EGOR IVANOVICH Application of Elliptic Functions to Questions about Functions Deviating the Least and the Most from Zero [Приложение эллиптических функций к вопросам о функциах, наименее и наиболее отклоняющихся от нуля], *зап. С-Петербургской Акад. наук,* **30,** St Petersburg 1877, *in:* [Zol32, вып. 2, S. 1-59].

[Zol77/2]   ZOLOTAREV, EGOR IVANOVICH *About the Scientific Work of O. I. Somov* [Об ученых трудах О. И. Сомова], St Petersburg 1877.

[Zol78]   ZOLOTAREV, EGOR IVANOVICH Sur l'Application des Fonctions Elliptiques aux Questions de Maxima et Minima, *Bull. de l'Acad. des Sciences de St. Pétersbourg,* 3ème *série,* **24,** St Petersburg 1878, *in:* [Zol32, вып. 1, S. 369-374].

[Zol32]   ZOLOTAREV, EGOR IVANOVICH Collected Works [Полное собрание сочинений], изд. АН СССР, Leningrad 1932.

[Zyg45]   ZYGMUND, A. Smooth Functions, *Duke Math. J.* **12** (1945), 47-76.

Index